NUCLEAR SILK ROAD

THE "KOREANIZATION" OF NUCLEAR POWER TECHNOLOGY

KIM BYUNG-KOO

طريق الحرير النووي

원 자 력 비 단 길

核之絲綢路

Copyright © 2011 Kim Byung-koo
All Rights Reserved.

ISBN: 1456422588
ISBN-13: 9781456422585

Library of Congress Control Number: 2010918280

TABLE OF CONTENTS

Nuclear Map of Korea i
Map of Daedeok Innopolis – Nuclear Technology Mecca ii
Nuclear Power Plants in the Silk Road Countries iii
Korean Nuclear Power Plants Chronology. iv

Korean Nomenclature v
Nuclear Acronyms vii

Foreword xvii
Acknowledgments xxi

Prologue xxv

Part I. Technology over Politics
Chapter 1: "Close down KAERI". 1
Chapter 2: The alternatives. 13
Chapter 3: Saviors 25
Chapter 4: The system designer 51
Chapter 5: Best mentor 61
Chapter 6: Aftershocks. 75
Chapter 7: Lessons learned the hard way. 83

Part II. Know-Hows and Know-Whys
Chapter 8: The formative years. 99
Chapter 9: Standardization 113
Chapter 10: Korean Pilgrims 129

Chapter 11: The Joint Design 145
Chapter 12: Growing pains 163
Chapter 13: Next-generation reactors 177
Chapter 14: Research reactor story 205

Epilogues:
Along the Nuclear Silk Road 213
Post–Fukushima Second Thoughts 219

Annex
I. Chronology of Nuclear Power Technology in Korea 229
II. Korean Nuclear Entities 235

Notes 249
Bibliography 255
Index 261

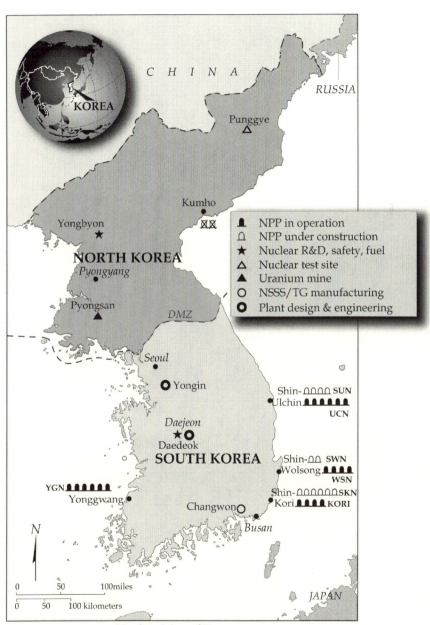

Nuclear Map of Korea

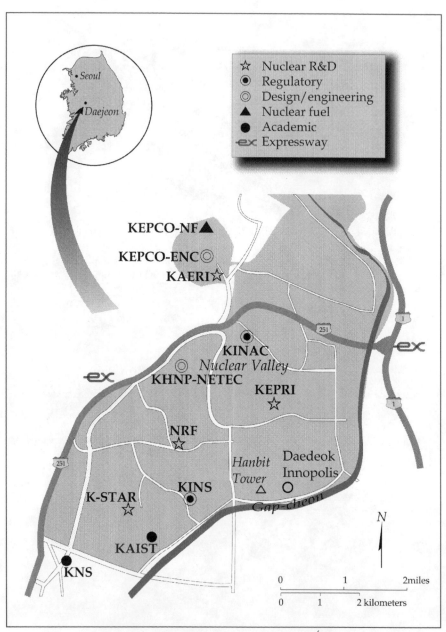

Map of Daedeok Innopolis - Nuclear Technology Mecca

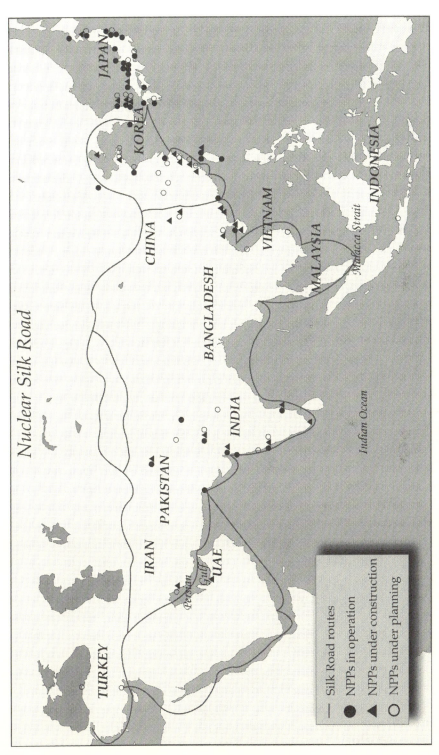

Nuclear Power Plants in the Silk Road Countries (2010)

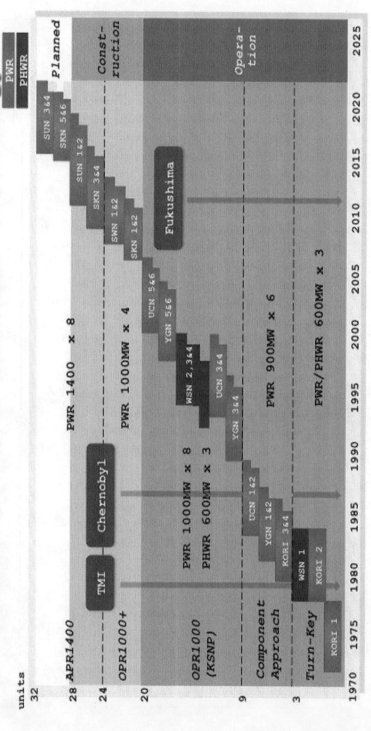

KOREAN NOMENCLATURE

Korean personal names can be spelled differently in English.
This book adopted the following for the sake of uniformity:
last name first, followed by given name with '-' between syllables as needed

> **Examples:** *Rhee Syngman, Kim Byung-koo*

There are thirty-two nuclear power plants in South Korea in operation or under construction, located in seven different nuclear sites. Three site names with *Shin-* (meaning 'new' in Korean) are located adjacent to the existing sites. The NPP sites and units (code name in parentheses) are spelled as:

Kori Units 1&2   (KORI 1&2)
Shin-Kori Units 1&2   (SKN 1&2)
Wolsong Units 1&2   (WSN 1&2)
Shin-Wolsong Units 1&2   (SWN 1&2)
Ulchin Units 1&2   (UCN 1&2)
Shin-Ulchin Units 1&2   (SUN 1&2)
Yonggwang Units 1&2   (YGN 1&2)

For the *Yonggwang 3&4* project which was central theme in this book, the acronym 'YGN' is used throughout.

Many references cited throughout this book are written in Korean and no English versions exist, the reference quotation is translated into English. When multiple contributors participated in the editing process, the Editor-in-chief's name is quoted as the author.

NUCLEAR ACRONYMS

ABB	Asea Brown Boveri
ABWR	Advanced Boiling Water Reactor developed by GE/Hitachi
ACRS	Advisory Committee on Reactor Safety
ADD	Agency for Defense Development, ROK Ministry of Defense
A/E	Architect-engineering, overall plant design of nuclear power plant
AEC	Atomic Energy Commission, ROK Prime Minister, the chairman
AECL	Atomic Energy of Canada Limited, Canadian crown corporation for nuclear research and power projects
AERI	Atomic Energy Research Institute, predecessor to KAERI (1959-1973)
AGR	Advanced Gas-cooled Reactor, originally developed in the UK
AIF	Atomic Industrial Forum
ALWR	Advanced Light Water Reactor, Generation III reactor concept developed in the US
APR1400	Power Reactor of 1,400 MWe, developed by KEPCO consortium
APWR	Advanced Pressurized Water Reactor, developed by Mitsubishi
AP1000	Advanced PWR of 1,000 MWe, developed by Westinghouse
ATLAS	Advanced Thermal-hydraulic Loop for Advanced Simulation, at KAERI

BNFL	British Nuclear Fuel Limited
BNPP	Braka Nuclear Power Plant in UAE
BWR	Boiling Water Reactor originally developed in US
CANDU	CANadian Deuterium Uranium, pressurized heavy water reactor developed by AECL
CANFLEX	CANDU Flexible, advanced CANDU fuel jointly developed by AECL & KAERI
CE (or C-E)	Combustion Engineering, US NSSS vendor company, later merged with Westinghouse (2000)
CERCA	*Compagnio pour l'Etude et la Realisation de Combustible Atomiques*, French nuclear fuel fabrication company
CESSAR	CE Standard Safety Analysis Report
CP	Construction Permit issued by the national regulatory before the first concrete is poured
CRT	Class Room Training
DC	Design Certificate, standard design approval by US-NRC on a generic NPP design
DPRK	Democratic People's Republic of Korea, otherwise known as North Korea
DUPIC	Direct Use of PWR fuel In CANDU, novel fuel cycle concept by KAERI
ENEC	Emirates Nuclear Energy Corporation, nuclear utility company of UAE
EPB	Economic Planning Board, ministry under ROK Prime Minister
EPGCC	Electric Power Group Cooperation Council, KEPCO consortium
EPR	European PWR Reactor, developed by Areva
FANR	Federal Authority for Nuclear Regulation, regulatory body in UAE
FSAR	Final Safety Analysis Report
FSE	Fluid Systems Engineering, a discipline of reactor system design

HANARO	High-flux Advanced Neutron Application Reactor, 30 MW multipurpose research reactor at KAERI
HTGR	High Temperature Gas-cooled Reactor originally developed in the US and Germany
IAEA	International Atomic Energy Agency, headquarters in Vienna, Austria
ICD	Initial Core Design, nuclear fuel software design
I&C	Instrumentation and Control, a discipline of reactor system design
INES	International Nuclear & Radiological Event Scale, from Level 1 to Level 7
INFCIRC	INFormation CIRcular, series of legally binding official IAEA publications
INPO	Institute of Nuclear Power Operators
INSAG	International Nuclear Safety Advisory Group in the IAEA
ITB	Invitation-to-Bid, formal tender document issued by the owner
JAERI	Japan Atomic Energy Research Institute
JD	Joint Design, contractually binding design process between licensor/licensee
JSCNOET	Joint Standing Committee on Nuclear & Other Energy Technologies, bilateralnuclear standing committee between the US and ROK
JSD	Joint System Design, form of JD for NSSS system design between KAERI and CE
KABAR	Korea Atomic Burns & Roe, the first architect-engineering company in Korea
KAERI	Korea Atomic Energy Research Institute; 'Atomic' was changed to 'Advanced' in 1980, then restored to 'Atomic' in 1989
KAIST	Korea Advanced Institute of Science & Technology, the highest science and engineering university in Korea, located in the Daedeok Innopolis, Daejeon

KECO	Korea Electric Company, predecessor to KEPCO until 1982
KEDO	Korean Peninsula Energy Development Organization
KEPCO	Korea Electric Power Corporation, restructured from KECO in 1982, sole electric utility company in Korea
KEPCO-E&C	KEPCO Engineering & Construction, new name for KOPEC
KEPCO-NF	KEPCO Nuclear Fuel, new name for KNFC
KEPIC	Korea Electric Power Industry Codes, industrial codes for NPP in Korea
KEPRI	Korea Electric Power Research Institute at Daedeok
KHIC	Korea Heavy Industries Corporation, later changed to Doosan Heavy Industries
KHNP	Korea Hydro & Nuclear Power, wholly owned subsidiary of KEPCO, owns and operates all NPPs, hydro plants, and pumped storage plants in Korea
KHNP-NETEC	KHNP Nuclear Engineering Technology Center at Daedeok
KINAC	Korea Institute of Nuclear Nonproliferation & Control
KINS	Korea Institute of Nuclear Safety, nuclear regulatory body
KIRAMS	Korea Institute of Radiological & Medical Sciences
KMA	Korea Military Academy
KMRR	Korea Multi-purpose Research Reactor
KNE	Korea Nuclear Engineering, later changed to 'KOPEC'
KNFC	Korea Nuclear Fuel Company, later changed to 'KNF'
KNFDI	Korea Nuclear Fuel Development Institute
KNGR	Korean Next Generation Reactor program
KNICS	Korea Nuclear Instrumentation & Control Systems
KNS	Korean Nuclear Society, an academic society

KOPEC	Korea Power Engineering Company, name changed to KEPCO-E&C
KRMC	Korea Radwaste Management Corporation
KRR-1/2	Korea Research Reactor, units 1 and 2
KSNP	Korean Standardized Nuclear Power Plant
KWU	Kraftwerk Union, part of Siemens, nuclear vendor company in Germany
LEU	Low Enriched Uranium fuel, below 20% enrichment
LWR	Light Water Rector
MDA	Mechanical Design & Analysis, a discipline in reactor system design
MER	Ministry of Energy & Resources, ROK government
MEST	Ministry of Education, Science & Technology, responsible for nuclear regulatory and research
MKE	Ministry of Knowledge Management, responsible for nuclear industries
MMIS	Man Machine Interface System, control room design of NPP
MOST	of Science & Technology, ROK government
MWe	Mega Watt electric output of NPP
MWt	Mega Watt thermal output of NPP
NPP	Nuclear Power Plant
NPT	Nuclear Nonproliferation Treaty
NRX	Nuclear Research Experimental, research reactor at Chalk River, Canada
NSC	Nuclear Safety Center, predecessor of KINS at KAERI
NSSS	Nuclear Steam Supply System
OAE	Office of Atomic Energy under MOST
OJP	On-the-job Participation
OJT	On-the-job Training
OPR	Optimized Power Reactor, new name for the KSNP reactor
PHWR	Pressurized Heavy Water Reactor

PNC	Power Reactor & Nuclear Fuel Cycle Corporation, Japan
PRSD	Power Reactor System Division at KAERI
PSAR	Preliminary Safety Analysis Report
PWR	Pressurized Water Reactor, one type of LWR
RCP	Reactor Coolant Pump
RE	Reactor Engineering, a discipline of reactor system design
SA	Safety Analysis, a discipline in reactor system design
SAGSI	Standing Advisory Group on Safeguards Implementation
S&L	Sargent & Lundy, architect-engineering firm in the US
SBO	Station Blackout, total loss of electricity in a nuclear plant under emergency
SD	System Design, the core technology in NSSS
SDA	Standard Design Approval, design certification of a generic design NPP
SDS	Safety Depressurization System
SKN	Shin-Kori Nuclear Power Plants
SMART	System-integrated Modular Advanced Reactor developed by KAERI
SMR	Small and Medium Reactor
SSAC	State System of Accountancy Control
SSAR	Standard Safety Analysis Report
SUN	Shin-Ulchin Nuclear Power Plants
SWN	Shin-Wolsong Nuclear Power Plants
SNU	Seoul National University
T/G	Turbine and Generator
TMI	Three Mile Island, NPP in Pennsylvania, USA, known for the accident in 1979
TRIGA	Training Research Isotope reactor developed by General Atomic, USA
TT	Technology Transfer, technology licensing agreement

UAE	United Arab Emirates
UCN	Ulchin Nuclear, NPPs built at Ulchin site in Korea
USDOE	US Department of Energy
USNRC	US Nuclear Regulatory Commission
VVER	PWR system design developed in Russia
W (or WEC)	Westinghouse Electric Corporation
WSN	Wolsong Nuclear, NPPs built at Wolsong site in Korea
YGN	Yonggwang Nuclear units 3&4 NPPs, the central theme to this book

The Lamp of the East

In the golden age of Asia
Korea was one of its lamp-bearers
And that lamp is waiting to be lighted once again
For the illumination in the East.

Rabindranath Tagore, 1929
[Bengali poet, the first Asian Nobel
Laureate in Literature, 1913]

東方의 등불
일찍이 아시아의 황금 시기에
빛나던 등불의 하나였던 코리아,
그 등불 다시 한 번 켜지는 날에
너는 동방의 밝은 빛이 되리라.

[주요한 역, *1929*]

FOREWORD

This book is intended for international readers who may want to better understand the South Korean nuclear industrial success story. The news of the UAE nuclear power plant contract to the Korean consortium in 2009 gave a compelling reason to the author to write the book. South Korea is a striking example of how the power of technology and the imagination of leaders have combined first to achieve sustainable development with the use of nuclear power to cover domestic electricity needs, then reaching out to the export market.

The Koreanization process of the nuclear power technology, from design, manufacturing, construction, operation & maintenance, fuel fabrication and building up of a safety regulatory infrastructure in a relatively short period is worthy of note. This book offers an insider's view of the localization process of Korean nuclear power technology. South Korea was fortunate to have its highest political leaders consistently supporting nuclear energy policies through their personal commitments to nuclear power for the energy resource poor country. Critical moments in the 1980s are described in detail when the national NPP technical self-reliance policy was put on a fast track with the construction of Yonggwang 3&4 Units. This became the benchmark model for the subsequent ten Korean Standardized Nuclear Plants built and operated in Korea. It all started with the localization leadership of the national nuclear research lab (KAERI) supplying the brain power, and the state owned electric utility (KEPCO) supplying the management oversight and necessary funding for the construction as well as for the technology transfer. They were able to work out a unique scheme to overcome shortages of experienced manpower, as well as the constraints in time and of budget that typically prevail in a developing country. Through repeat projects of the standardized NPPs in the 1990s, the Korean entities learnt the know-hows from their US mentors. Then they moved on to their own by learning the know-whys to the new generation of advanced pressurized water reactor technology with improved safety and economics for the competitive world market.

I believe the author has unique qualifications to tell the story. As the project manager of the very first nuclear reactor system design during the crucial period in the 1980s, he was one of the key player in the Korean self-reliance drive. He then moved on to serving as an international civil servant at the IAEA. As a director of Technical Cooperation assisting many developing countries that wish to acquire nuclear power technologies, he gained broad experience and perspectives of the needs of developing countries.

Korean Peninsula today exemplifies an extreme nuclear dichotomy. Since the 1970s, South Korea is pursuing a national drive for the peaceful uses of nuclear energy, focusing first of all on NPP construction and associated technology of self-reliance. North Korea went for the opposite side of the nuclear coin and developed nuclear weapons capable of mass destruction. Today the Northern half of the Peninsula is left in darkness and isolation aggravating regional insecurity and undermining the well-being of its own people.

Nuclear renaissance is more visible in Asia than in any other area today. China, India and UAE connect the Far East (Japan and Korea) with the Middle East covering the same route as once the ancient Silk Road. The world economy will continue to be driven by the emerging economies demanding ever more clean electricity and outpacing the developed economies by a factor of four in the foreseeable future. The international nuclear community must assist and facilitate the Asian nuclear renaissance and help ensure application of global standards of nuclear safety, security and non-proliferation – in the sphere of research, in the

Acknowledgments

industrial sector, in trade and at the highest policy making levels. I congratulate the author of this book and his fellow Korean trailblazers on the dedicated and outstanding professional work they have performed. The lessons learned from the Korean experience will be useful to all like-minded readers of this book.

Hans Blix

ACKNOWLEDGMENTS

This book is dedicated to the unsung heroes of the Koreanization of nuclear power technology who gave their life-time passion to make the impossible dream come true. During the course of this writing, I was humbled to discover thousands of names and their contributions from research institutes, industries, regulatory boards, academia and from government ministries over the past half-century; only a few are named in this book.

I am grateful that I was privileged to serve in the capacity of being able to manage one of the key elements of NPP technology in the reactor system design with a team of engineers and technocrats from both sides of the Pacific. Moving on to the international arena at the IAEA, I was often asked and puzzled by the question "Can you please explain how you Koreans did it?" by my friends from many different parts of the world who were genuinely interested in setting up their own nuclear power infrastructure. My sincere wish is that this book will serve to satisfy some of their inquisitiveness and help them realize it was no "miracle" but plenty of "sweat and tears," with countless agonies and in-fights. My country is blessed with human resources that took part in the Koreanization process, none less than the consistent support from the very top of the government.

In collecting the raw data on the history of what did happen in Korea, I was fortunate to be living at my old home, *Casablanca* (meaning 'white house' in Spanish), located at the heart of the nuclear technology Mecca, Daedeok Innopolis, right after my retirement from the IAEA. From my proximity to most of the major nuclear centers and my old acquaintances, I was immensely benefited to discover accurate accounts of how the technology self-reliance program proceeded through hundreds of internal reports and documentations. In addition, live interviews with surviving key players in Korea, and in the US were rewarding experiences for me to recollect the past in proper perspective. My whole purpose of writing this book is to bring out the correct and complete picture (in the good sense of the IAEA safeguards) of the Koreanization process without losing the reader's interest. After all, this is intended to be a nonfiction documentary of the history of nuclear power for general readership, and is not intended to be a technical book for nuclear specialists. Despite all my efforts to be as correct and complete as possible, there may be shortcomings and incorrectness, for which no one but I will be solely responsible. I would appreciate information about any errors to be sent to me (bkkim9@gmail.com), so those could be duly reflected in the next edition.

I am immensely indebted to my old friends and colleagues at KAERI, KINS, KEPCO, KEPCO-E&C, Doosan, Westinghouse (old CE), and the ROK Ministries for providing invaluable input and advice. Special thanks are due to Yoon Young-ku, Cha Jong-hee, Lee Hae, Rim Chang-saeng, Chang In-soon, Kim Duck-seung, Kim Si-hwan, Choi Chang-oong, Han Ki-in, Lee Byung-ryung, Lee Ik-hwan, Choi Suhn, Sohn Gap-heon, Cho Kwang-hee, Park Jung-ki, Lim Han-kwae, Rieh Chong-hun, Jeon Jae-poong, Park Yong-taek, Roh Eun-rae, Suh Suk-chun, Chang Soon-heung, Kim Jong-hyun, Lee Seung-hyuk, Lee Seung-koo, Kim Se-jong and Chung Chung-woon for their significant personal input and encouragement. From the US side, Jim Crawford, Tom Natan, Regis Matzie, Mark Crump, and Robert S. Lee (deep condolences, who passed away during this writing) provided the balancing viewpoints from the mentor's perspective. Also, special thanks are due to Ken Rohde, who introduced me to the new world of self-publishing from CreateSpace, an Amazon.com company, which saved me from finding a literary agent. But, most of all, I would not be able to finish this book without Han Pil-soon, Kang Bak-kwang, Jim Veirs, Shim Chang-saeng, Lee Chang-kun, and Nam Jang-soo, who personally gave their time and effort to screen and correct my manuscripts. They provided me with a continuous source of inspiration throughout. Family support from my uncles, CJ and YJ, and of course my dearest wife, Lucia, was also crucial for me to keep on writing and kept me from giving up.

PROLOGUE

The purpose of *Nuclear Silk Road* is to introduce to the outside world, from an insider's view, the history of development of the Korean nuclear power program as it happened during the formative twenty years between 1975 and 1995. The year 1975 was when South Korea formally ended its short-lived fuel cycle program, and the Nuclear Nonproliferation Treaty (NPT) was entered into full force, signaling the beginning of a new era of peaceful use of nuclear energy. The year 1995 was also historic in that the first original Korean Standardized Nuclear Power (KSNP), Yonggwang 3&4 (YGN), went into commercial operation, together with the achievement of the technical self-reliance goal, allowing all nuclear power projects to be retransferred from KAERI to the industrial sector.

However, it is necessary to understand the background of the nuclear program, which started from the late 1950s under the first President Rhee Syngman. He was a statesman and lifetime independence movement fighter from the Japanese occupation, who turned into a fierce anti-communist fighter since the first Republic in 1948. President Rhee had a special vision for nuclear energy while the country was still recovering from the devastating Korean War (1950-1953), creating the basic nuclear infrastructure in the government. A series of Atomic Energy Laws was promulgated, and the Office of Atomic Energy was established directly under the President, together with the Atomic Energy Research Institute (which later became KAERI) in 1959. Moreover, a small research reactor, TRIGA Mark-II, was constructed at KAERI in the same year. Considering the economic state of Korea at that time, with a per capita income of less than one hundred dollars, it was a visionary move by the old statesman. Nuclear engineering departments were established in 1958/1959 at the nation's most prestigious universities (Hanyang University and Seoul National University, respectively), attracting the brightest among high school graduates. The first two decades were the seeding period of nuclear basic science in Korea, centered around KAERI's newly built TRIGA research reactors, breeding the most basic requirements in nuclear power: the development of human resources.

The decision to build the nation's first nuclear power plant, Kori Unit 1, was made in 1968; it was a total turnkey project from Westinghouse, and went into commercial operation in 1978. When the fuel cycle program was scrapped, creating further confusion and demoralization of staff, KAERI began to get involved in commercial nuclear-power-related projects. Then it came to the serious business of challenging the localization of core nuclear power technology in the 1980s, however, the South Korean infrastructure had every reason to fail: lack of experienced manpower, lack of budget, and no time to develop from scratch, just to name a few. The "Korean miracle" story was about to hatch when the brain power of KAERI came together with the management power of KEPCO for the first time. It all started with the creation of an architect-engineering company (which later became KOPEC) and the feasibility study of nuclear fuel localization projects in the late 1970s. Their initial target was CANDU and PWR-type nuclear fuels. The first attempt to challenge the full-scope nuclear power technical self-reliance by the Korean scientists and engineers did not come without another historic episode in the 1980s, imbedded within the political turmoil in the country.

This book is divided into two parts. *Part I. Technology over Politics* focuses on the unique history of the nuclear power infrastructure-building process centered around the YGN project inception in the 1980s, when the technical self-reliance goal was pushed, simultaneously with the new NPP construction project. A dramatic sequence of events unfolded from the "close down KAERI" episode to the selection of the technology partner, CE, then to the totally unprepared-for aftershocks, and the lessons learned the hard way, including the anti-nuclear movements and the North Korean KEDO project. What appeared to be a "miracle" at the time to allow technical recommendations dictating the final outcome of the normally political decision for selecting the technology partner is unveiled in the *Part I*.. *Part II. Know-hows and Know-whys* focuses on how the process of technical self-reliance was implemented once the political dust settled down. The preceding formative years before the YGN project are introduced to help readers understand the fertile soil under the standardization study project. A detailed description of how the technology transfer from Combustion Engineering with the Joint Design approach is followed by the growing pains of setting up the national regulatory system to assure nuclear safety, together with the YGN project. Moving forward with the fleet of KSNP constructions, the next-generation reactor development effort was vigorously pursued to the Generation III reactor APR1400, which was exported to UAE. It is of historic coincidence that the nuclear renaissance in the Asian countries, notably in China, India, and UAE, will inevitably connect the ancient Silk Road trails from Korea in the Far East to the Middle Eastern countries on freight carriers and super jumbos instead of camels and horses.

A tragic nuclear accident hit Fukushima NPPs in northeastern Japan following the record high earthquake and subsequent tsunami on March 11, 2011, as this book was about to go to print. It resulted in a significant amout of radiation release to the environment, only second to Chernobyl in history (IAEA International Nuclear and Radiological Event Scale 7), as well as the complete loss of four NPP units at Fukushima Daiichi. A new epilogue was added at the end of this book, "Post-Fukushima Second Thoughts", to reflect the nuclear safety issues and the great nuclear dilemma it may bring. The author's wish is that the world nuclear community will learn valuable lessons for safer nuclear power in the future.

PART I. TECHNOLOGY OVER POLITICS

CHAPTER 1:
"CLOSE DOWN KAERI"

Minister's Order

It happened in the morning of the second week of September 1980. Kang Bak-kwang (it is the Korean custom to precede the given name with the family name), director general of the Atomic Energy Development Bureau(AEDB, responsible for supporting all national nuclear R&D activities) of the Ministry of Science & Technology (MOST), was summoned to the minister's office on the nineteenth floor of the government building in Seoul. Professor Lee Cheong-o was recently appointed as the first minister of MOST just a few days before, under the newly formed Chun Doo-hwan's 5th Republic (President Chun was inaugurated on August 27, 1980). Director General Kang had personally known the minister before from helping the then-professor in the mechanical engineering department at KAIST to get a sabbatical leave to France for three years. Now, his boss, Minister Lee, was a scholarly man, having returned from France recently, and hopefully seeking advice from Kang on some nuclear matters, or so Kang thought. Contrary to his expectations, however, Minister Lee gave a straight order, a simple statement saying, "You proceed to close down KAERI (Korea Atomic Energy Research Institute)." Kang, shocked and embarrassed to ask any questions right away, only said "I will review it, Minister," and then walked out of the minister's office. What made Kang more embarrassed was that he was also newly appointed to the current position only a month ago from the head of the International Technical Cooperation Bureau.

Being a seasoned technocrat himself, Kang immediately sensed there must have been a grave agenda behind the minister's order to close down KAERI, the oldest and largest national research institute, and the one and only nuclear research center in Korea. His guess was that the selection of Professor Lee for the cabinet post must have had something to do with carrying out the closure order. Maybe the newly-chosen Minister, a close confidant of President Chun, graduating from the thirteenth class of the Korea Military Academy (KMA); thus maybe he was given the single mission to carry out something drastic in the nuclear sector. It did not seem feasible that the new minister had some inside knowledge about KAERI's status to conclude such a terminal measure, as Kang had yet to report to the minister on the KAERI issue by then. Further, he may not have been familiar with domestic science and technology matters in general, and nuclear KAERI matters in particular, due to his three-year absence from Korea until just a few months prior.

A few days later the minister asked Kang again about the status of the closure. Better prepared this time, Kang answered, "It is not so simple, sir, to close down KAERI for three reasons: (1) an historic reason being that KAERI was created twenty years ago by the first president of the republic, Rhee Syngman, following Eisenhower's 'Atoms for Peace' declaration to open the peaceful uses of atomic energy in Korea; (2) KAERI was founded under a special legislation passed by the national assembly two decades ago, and it would require a consent of the national assembly to close it down by abolishing the KAERI Establishment Law; and (3) Korea would need to develop a nuclear energy option for energy independence from fossil fuel, which is one hundred percent imported. Thus, closing down the sole atomic research institute means we are giving up nuclear power for the future." It must have taken some guts on Kang's part to talk back in defiance to his minister's order, never mind how logical his arguments may have been. Kang went on, saying to the minister, "You must convince me first of the reasons why we must close down KAERI, sir."

Kang was no ordinary Director General in speaking up to his boss. A graduate of Waterloo University in Canada, with a chemical engineering PhD, he repatriated to Korea in 1974 to become the Directorate for Nuclear Safeguards at MOST, as one of the first overseas recruits to a director general's rank (the youngest, at thirty-four). He was able to stand up for his convictions of what was good for the country beyond any ministerial turf fights, partly due to his family support. Kim Young-june, president of KEPCO (Korea Electric Power Corporation) in the 1970s and a close confidant of President Park Chung-hee, was Kang's father-in-law, and provided guidance and liaising between MOST and the Ministry of Energy & Resources (MER, to which KEPCO reports), between which there was a traditional rivalry on nuclear matters.

The minister, taken back by Kang's contentious defiance, reluctantly came up with the following two arguments for closing down KAERI. First, Korea was building its first NPP and KAERI was not contributing at all in assuring safety against possible nuclear accidents. KAERI did not have enough technical manpower on nuclear safety, so Korea would have to depend entirely on foreign technical supports. Second, KAERI should have focused on power reactor and nuclear fuel research programs, but, to the contrary, their focus was only on non-power areas like agriculture and radioisotope research. If these were the real reasons, Kang felt, it could justify further beefing up KAERI instead of closing it down, thus leaving the strong impression that these were just the minister's excuses while the real reason was still hidden. The minister even said, "What are the Atomic Energy Commissioners doing? Why not abolish the AEC office?" There were two AEC commissioners in the ministry at that time in accordance with the Atomic Energy Law. This led to the temporary relocation of the two AEC commissioners' offices to a KAERI site located in a suburb of Seoul. Frustrated, further aggravated by Kang's protest, and not being able to tell the real reason, the minister spoke almost in desperation, "Then you come up with alternatives!" Kang had a strong suspicion that behind the minister's order lay a confidential diplomatic bargain between Seoul and Washington on the dismantlement of the Korean nuclear research and missile program in lieu of winning US diplomatic recognition of the new regime.

Kang responded right away: "Sir, you mentioned two problems: NPP safety on one hand and focusing nuclear R&D on NPP on the other." He thought they could find solutions for both problems and promised to come up with a new plan and report to the minister within a month. For the first issue of NPP safety, Kang organized a confidential study group within KAERI headed by Lee Sang-hoon (then Safety Section head, later serving as the first president of KINS). A plan on the establishment of the Nuclear Safety Center was reported to Minister Lee within a month, as promised. For the second issue of focusing nuclear R&D on NPP, which involved the restructuring of KAERI, Kang reported that such an idea must be transmitted to the ongoing study of the special task force on restructuring all national research organizations. Kang felt that Minister Lee, after listening to the Nuclear Safety Center plan, seemed to have moved away from his strong attitude for closing down KAERI and secretly changed his mind to take actions to save KAERI.[1]

This episode is surfacing for the first time after some thirty years. In fact, something "good from evil, or opportunities from crisis" has developed during the next decades that finally led to the NPP export to the UAE in 2009, which can be traced back to this dark period in Korean nuclear history. It is the essence of this book: how the "Koreanization" of nuclear power technology will unfold, and KAERI's role change in the 1980s in particular during this dramatic episode. Understanding of the socio-political, diplomatic situations surrounding the Korean Peninsula in early 1980 is required to uncover the reasons.[2]

First two decades

It is important to understand the history of KAERI during the first two decades leading up to its "close down" episode of 1980. The Atomic Energy Research Institute (AERI, the predecessor of KAERI) was founded in 1959 by Rhee Syngman, the first president of the first republic, inspired by Eisenhower's "Atoms for Peace," as were many other nations during that era. Building up a national nuclear research center from virtually no technical infrastructure to speak of at that time, their focus was on manpower development in basic sciences like nuclear physics, radiation chemistry, and life sciences. All AERI staff members were civil servants of the "nuclear research officer" with unprecedented privileges and salaries much higher than ordinary civil servants. The construction and operation of two small-scale research reactors (TRIGA Mark-II and TRIGA Mark-III) located side-by-side at Taenung (adjacent to the Engineering College of Seoul National University campus) just outside of Seoul led to enough radioisotope productions for medical and agricultural research. Inspired by this modern research center in full swing, the brightest young students were enrolled at the newly opened Nuclear Engineering Departments at Hanyang University and SNU from 1958. (This was earmarked by the highest college entrance exam cutline scores for the nuclear engineering department for several years). After a decade of AERI operation, it was beginning to lose its original glory with the fragmentation of the research institute into Radiation Medical Institute and Radiation Agricultural Institute. When national planning for the first nuclear power plant construction came about in the late 1960s, AERI played only a minor role in advising the government. Instead, the state-run electric utility company KEPCO took the leading role in planning and managing the new NPP construction projects (starting with the Kori Unit 1, Westinghouse 600 MWe PWR turnkey project) in the early 1970s. It was a detrimental blow to AERI, once a proud premier nuclear entity in Korea.

Come 1973, AERI was restructured from a civil service institute to a government-sponsored autonomous research institute in line with the changing times. The official name of the institute was also changed to add "Korea" at the front, to be called "KAERI." The first president of the newly restructured KAERI was Yoon Young-ku, a Brown University-educated chemist with nuclear materials experience from Argonne National Laboratory. Yoon's first task was to rebuild KAERI to be more focused on nuclear power and fuel cycle technologies in line with the onset of the nation's first nuclear power plant Kori Unit 1. It was a drastic change of mission from the existing nuclear basic science mentality to more engineering oriented nuclear power technologies. He was so much ahead of his time to the vision of energy independent Korea with its nuclear power program and associated fuel cycle technology at its infancy.[3] He began rigorous recruiting campaigns from overseas Koreans with nuclear power experience. His first successful recruit was Lee Hae from Westinghouse Nuclear Research Center in 1974. Graduating from Penn State with a PhD in mechanical engineering, Lee had a flow-induced vibration research background at the Westinghouse Research Center, Pittsburgh. Lee was the first technical staff member to join KAERI with solid engineering experience of PWR-type NPP. Kim Chong-joo, vice president of KEPCO, was also recruited to take up the newly created vice president of engineering position at KAERI. He founded the very first architect-engineering company, KABAR, in 1975, together with Burns & Roe of the USA, which subsequently became KOPEC in 1982. It was the first attempt of a national nuclear laboratory to get into the commercial NPP business in architect-engineering. The year 1975 should be remembered as the first year the national nuclear lab embarked on its long journey to commercial NPP technology, thanks to its pioneering leaders like Yoon, Kim, and Lee. About twenty overseas Korean engineers were recruited under Yoon's leadership to join the newly reborn KAERI (and the Korean nuclear power technologies) at its infancy in the 1970s. Many overseas recruits made significant contributions to the Korean and international nuclear scene: Rim Chang-saeng (MIT graduate, PhD in nuclear engineering) and Kim Seong-yun (NYU, PhD in nuclear engineering) both served as KAERI presidents in the 1990s; Juhn Poong-eil (Carnegie-Mellon, PhD in nuclear engineering) and Kim Byung-koo (Caltech, PhD in applied mechanics) both served as technical directors at the IAEA, Vienna.

One other unfulfilled but meaningful episode of this period involved the CANDU-type heavy water reactor localization projects led by KAERI. Standardized CANDU NPPs were being built in Canada, mainly in Ontario, plus several export projects were also being built in Argentina

(Embalse NPP) and Romania (Cernavoda NPP) in the early 1970s. CANDU offered definite advantages for developing countries like Korea in that, in those days, the fuel required was natural uranium, thus eliminating the need for enrichment, plus there was a perceived ease of localizing major components due to its inherent lower pressure and temperature conditions. The only handicap was the need for heavy water in large volumes, which was very expensive and considered sensitive material. The first feasibility study was conducted jointly by KAERI and KEPCO in 1973. Four high-level expert members were sent to Canada to consider a CANDU option; Hyun Kyung-ho and Cha Jong-hee (both of whom later served as KAERI presidents) from KAERI, and Moon Hee-sung and Roh Eun-rae from KEPCO, made important policy recommendations following their trip to introduce the first CANDU NPP at Wolsong.[4] A more serious six-month feasibility study of introducing a large fleet of CANDUs with localization of fuel, reactor equipment, and heavy water with a full scope of technology transfer was conducted jointly by KAERI and AECL in 1978. The comprehensive four-volume Joint Canada-Korea Study on Korean CANDU program, known as the JCKS report, was issued by Lee Hae and George Pon, both vice presidents and the project managers of KAERI and AECL respectively. They were ambitious in recommending 4x900 MWe CANDUs (referenced to Bruce A NPP) and an engineering development laboratory (referenced to AECL Sheridan Park Engineering Laboratory) to the Korean government.[5] After many twists and turns of history, 4x600 MWe-size CANDU plants were finally built at Wolsong in the 1990s with less localization contents than the JCKS proposed, as the Korean standardization with the self-reliance program was already decided with PWR-type NPP from the YGN project. A proposal to construct an engineering development laboratory for reactor equipment localization was partially realized with the construction of the hot test loop at KAERI during the Wolsong fuel localization project (as described in Chapter 3).[6] KAERI's intention to move into commercial NPPs with technical self-reliance had to wait eight more years for the YGN system design participation, not for CANDU (as described in Chapter 4).

French connection

In the early 1970s, a special laboratory group was formulated at KAERI to start on fuel cycle research projects. This was expanded and later referred to with the nickname "French Loan Projects," as a significant amount of French government loan was arranged to build fuel cycle facilities, including a pilot scale fuel cycle plant. Yet another reactor engineering team was given the task to design a material testing reactor, possibly with the Canadian NRX type in mind. Both projects were of special priority, with tight security requirements rather unusual in KAERI's normal academic atmosphere. These two special KAERI projects progressed to the detail design stage when the Indian nuclear test was conducted in 1974. International nonproliferation concerns were mounting to force cancellation of both projects just in time for Korea's entry into force of the Nuclear Nonproliferation Treaty (NPT), and subsequent safeguards arrangements by the IAEA at the end of 1975. History has it that the ROK national assembly ratified the NPT, leading to the entry into force, with the full scope comprehensive safeguards agreement with the IAEA (INFCIRC/140) on November 14, 1975. The IAEA inspectors were given full access to all nuclear facilities, confirming the absence of undeclared nuclear materials, including restructured fuel cycle activities at the KAERI Daedeok site.

A new spin-off entity from KAERI, the Korea Nuclear Fuel Development Institute (KNFDI), was founded in 1976 to focus on the follow-up French Loan Projects at Daedeok Science Town, one hundred and sixty kilometers south of Seoul. Its mission was to convert the original back-end fuel cycle project into the front-end fuel cycle activities in support of the nation's nuclear power program. Later on, the fuel cycle pilot plant construction project was reformulated for fuel fabrication, uranium conversion plants, and post-irradiation examination facility in the 1980s; and the material testing reactor construction was also reformulated to a multi-purpose research reactor, HANARO (30 MW open-pool type with low enriched uranium, well within the nonproliferation guideline), in the 1990s. This 1970s episode at KAERI sealed South Korea's only attempt at the sensitive fuel cycle technology in history, leading to the

Joint Declaration of Denuclearization of the Korean Peninsula in 1992, which forbade any reprocessing and enrichment facilities in the Korean Peninsula. The short-lived KNFDI was re-merged with KAERI in 1980 as an alternative solution to the "close down" episode, only four years after its creation. However, KNFDI left an important mark on the nuclear fuel localization program triggering the subsequent nuclear power technology self-reliance drive of the 1980s. Already the nation's first NPP Kori Unit 1 (600 MWe PWR supplied by Westinghouse) was in commercial operation since 1978, and the second NPP Wolsong Unit 1 (600 MWe CANDU-PHWR supplied by AECL) was under construction for grid connection by 1983. Since both units were burning nuclear fuels imported from the US and Canada, respectively, with more units being planned, rationale and justification for fuel localization projects of both PWR and PHWR types made economic sense for commercial-size fuel fabrication plants.

KNFDI issued two major policy recommendation reports on two types of nuclear fuels (CANDU-PHWR and PWR) to MOST in July 1980 with detailed techno-economic feasibility studies, just a few months prior to its reintegration with KAERI. It is important to note that the character of KNFDI, being the sole fuel development institute with full-scale commercial production in mind, was a drastic departure from any other national research labs in Korea at that time. The first priority was assigned to CANDU-PHWR fuel over PWR fuels for its simplicity and no need for enrichment service (crucial foreign dependence). It was deemed possible to develop full fuel design, in-core analysis, and fabrication capabilities within KNFDI without relying on technology transfers from AECL of Canada from the beginning (except for verification testing, if needed). The project code was named "W Project" for the time being. A five-year fuel development and demonstration program (1981-1986) was submitted to MOST for approval. Government funding approval came for the initial phase of two years, starting in 1981 after KNFDI's integration back to KAERI. It was the largest ever multi-year national project at KAERI, drawing on more than half of the total KAERI budget and manpower. Already the pilot scale uranium conversion and fabrication plants (annual ten ton-U capacity, later expanded to one hundred ton-U capacity for Wolsong Unit 1 full core supply) were being commissioned at the Daedeok site by 1981, assisted by French fuel fabrication company CERCA.[7]

More significant effort on KNFDI's part was put in strategic planning for PWR-type nuclear fuel. It was clear in the 1970s that PWR-type fuel demand would be rising as the same Westinghouse 600 MWe Kori Unit 2 was under construction with more PWRs in the planning stage. Thus, the PWR fuel localization plan submitted by KNFDI received favorable review by MOST as well as the Ministry of Energy & Resources (MER), which oversees KEPCO, the nuclear utility company. Creation of a new joint venture company with a foreign fuel company for fuel design and fabrication was proposed in 1981, which led to the creation of the Korea Nuclear Fuel Company (KNFC) in 1982 by KEPCO and KAERI. The site of the new KNFC fuel fabrication plant was earmarked initially near one of the existing NPP sites, but was later changed to the current site immediately adjacent to KAERI at Daedeok campus. Contrary to KNFDI's original plan, however, the major controlling share of KNFC went to KEPCO for its ability to finance the capital.[8]

It was the beginning of a shift of the center of gravity in nuclear research in Korea, not only in geographic locations from Taenung in Seoul to Daedeok Science Town in Daejeon, but also the basic character of the national nuclear laboratory. It was about to change from basic science to directly link to commercial nuclear fuel supply at the beginning, then maturing into nuclear reactor system design for power reactors in the late 1980s. The fuel-production-oriented KNFDI, aimed at supplying full-scale nuclear fuels for the nation's emerging commercial NPPs, lasted only four years (1976-1980). However, it played a historic role in erasing the memory of fuel cycle projects from KAERI's name to prepare for the coming decade of commercial projects at KAERI. KNFDI may have ended up being a lost child in the Korean nuclear scene, dictated by the international nonproliferation concerns since the Indian nuclear explosion of 1974. However, it produced the strategic master plans for both CANDU-PHWR and PWR-type nuclear fuel localization for government approval just prior to KAERI's shutdown episode of December, 1980. These master plans were fully implemented in the 1980s and 1990s by KNFC, the joint daughter company of KEPCO and KAERI.[9]

CHAPTER 2:
THE ALTERNATIVES

Desperate deliberation went on at the Director General Kang's office to come up with alternative ideas to closing down KAERI in the late 1980, when he received strange and confidential US visitors at his office. One identified himself as an US Army officer from the Far East Nuclear Strategic Forces, another as a Harvard University professor, escorted by US Embassy Science Officer Robert Stella. They were probing Kang on the KAERI closure progress, giving examples of the radical closure of a nuclear research center in Taiwan. (The Taiwanese government physically demolished the Taiwan Research Reactor, in operation for only two years, near Taipei in 1976, after the Indian explosion. The reactor was the same type as the Canadian NRX reactor). Kang argued that any back-end fuel cycle-related activities were totally scrapped in Korea since 1975 but legitimate peaceful use of nuclear power R&D activities must be allowed to carry on as Korea had already had its first NPP Kori Unit 1 in full operation since 1978. He somehow convinced the visitors that a new national nuclear R&D program would be provided in three months time demonstrating Korea's intention for strictly peaceful use of nuclear power.[1]

The new KAERI blueprint was developed in three months time in total secrecy - only very few knew the preparation work at MOST and KAERI. In the meantime, Minister Lee called on his most confided friend, Han Pil-soon (then-vice president of ADD, soon to become vice president of KAERI), in early 1982, and carried out secret work with him to make his own KAERI blueprint. Minister Lee came down to Daejeon almost every weekend for a while to work with Han at an unobtrusive place called Bangsan Inn, taking careful precaution not to be noticed by anyone. Nam Jang-soo of KAERI did secretarial work for them. Han and Nam had premonition that Minister Lee was preparing the briefing materials for the Blue House. The blueprint report contained the KAERI restructuring plans to move into commercial nuclear power technology, starting from a CANDU fuel localization project for Wolsong NPP, which was a drastic departure from the basic research of the past. History has proven that the major restructuring of KAERI based on the blueprint program developed at this critical period was the very backbone of Korea's success story in nuclear power in the next thirty years, and KAERI ended up playing the most crucial role. In other words, the US had to acknowledge the Korean rationale on peaceful nuclear power and the need for safety research as long as the Korean side assured the US of nonproliferation concerns with continued transparency. The alternatives contained three elements: 1) remove the key word "Atomic" from KAERI's title, 2) relocate the KAERI campus entirely from Seoul to Daejeon, and 3) create a new Nuclear Safety Center at KAERI.[2]

From "Atomic" to "Advanced"

All official historical archive publications mention this episode as KAERI simply being renamed the "Korea **Advanced** Energy Research Institute" from the original "Korea **Atomic** Energy Research Institute" in 1980 as part of the nationwide national institutes' restructuring and integration drive under the new military government of Chun Doo-hwan. The real reason for the disappearance of the word '**Atomic**' from the KAERI title had to wait for the next thirty years to be revealed.

An emergency Board meeting was summoned on December 19, 1980, passing a new resolution for changing the twenty-year old title of the institute to the Korea Advanced Energy Research Institute, now merged with KNFDI at Daedeok. In fact, the idea of switching from "Atomic" to "Advanced" was rather unique in that one may argue the validity of KAERI's closure to indicate no more nuclear R&D at the institute, on the surface. Koreans in general consider their names (or the title of an entity) in more symbolic way. This sudden disappearance of the key word "Atomic ('원자력' in Korean) generated profoundly uneasy feelings among the Korean scientific community in general, and the KAERI staff in particular. However, no one dared to ask, "Why?" as the country was in the midst of a martial law period at that time. Cha Jong-hee, then-president of KAERI in 1980, recalls in his personal memoir how desperate he felt when MOST Director General Baik Young-hak told him to remove "Atomic" from the KAERI title. Cha had no choice but to accept the name change in Korean, but came up with the idea of using "Advanced Energy" instead of "Atomic Energy", leaving the English acronym "KAERI" unchanged.[3]

In Korean, the new name "Korea Advanced Energy Research Institute" does not reflect exact meaning of the "Advanced Energy," but simply leaves ambiguity as to what the "Advanced Energy" really means. This created confusion in the other existing national energy research institute (Korea Energy Research Institute, 한국종합에너지연구소), under the Ministry of Energy and Resources, with a similar Korean name, forcing them to also change to "Korea Institute of Energy and Resources" (한국동력자원연구소) in 1981 for non-nuclear energy research. The name episode finally ended in 1989 when the old Korea Atomic Energy Research Institute (also 'KAERI') was formally reinstated, giving "Atomic" back its proper place. By then KAERI was in full force with commercial nuclear power plant projects (in nuclear steam supply system design and initial core design), leaving behind the old legacy of sensitive fuel cycle activities of the 1970s.

The decision to change the name from "Atomic" to "Advanced" brought immediate financial pains to KAERI as the Economic Planning Board, the national budgeting agency, began to cut KAERI project budgets in 1981. Government bureaucrats found easy excuses to cut the budget for KAERI, as it was advertised to no longer be in nuclear research. At one point in 1981, KAERI's bank account didn't have enough funds to pay even the basic staff salaries (as the national nuclear research institute, this happened only once in its entire fifty year history). The pain and agony of searching for meaningful alternatives was felt by everyone at KAERI as the uncertainty of their next salary payment became real. Staff morale was down to the bottom, and some able technical staff members did resign and find other jobs at university teaching posts.

Move to Daedeok

Another key element in the alternatives to the KAERI closure was to physically relocate the KAERI campus from its old Taenung site (a northeastern suburb of Seoul, already becoming too close to the population center in the 1980s) to the newly developed Daedeok Science Town near Daejeon south of Seoul. Combined with the name change, the physical relocation idea provided sufficient rationale and confidence to the high officials that indeed the old KAERI to be closed down for good, leaving impressions to outsiders that maybe no more nuclear R&D would be carried out in Korea. It was indeed good fortune for Korea to manage and survive this critical period in nuclear history as the main missions of KAERI changed drastically since 1981 to the commercial NPP technology. The "Koreanization" of nuclear power technology would not have been conceived without this turn of events in the early 1980s, creating a classic example of turning crisis into opportunities, as the subsequent chapters in this book will reveal.

Two main difficulties remained before the actual relocation to Daedeok could start. One was the phase-out and restructuring of the old fuel cycle activities; the other was the new construction of a research reactor to replace the two existing TRIGA reactors at Taenung campus, both to assure a proliferation-resistant nature. The first task was how to reprogram the "French Loan Project" from the original reprocessing pilot plant scheme into more nonproliferation-assuring fuel cycle activities. The alternate program to replace the reprocessing task was reached in 1976 among the KAERI management and the MOST officials to construct pilot uranium refining and conversion plants, together with a pilot fuel fabrication plant, all at the new Daedeok campus, while retaining the French loan amount with new targets. These new front-end fuel cycle plants were built in the late 1970s with French assistance, notably from the CERCA company. They played a crucial role in the early 1980s when KAERI began its new mission of nuclear fuel localization, starting with the CANDU-type fuel for Wolsong NPP. More significantly, KAERI moved down to Daedeok to build a new research reactor to replace the ailing TRIGA reactors, but also to assure a nonproliferation nature of the new research reactor. MOST assured KAERI of its intention to fund this ambitious multi-year construction project before the decision to move down to Daedeok was made. This turned out to be a much longer and expensive project, but finally a new multi-purpose research reactor HANARO (High-flux Advanced Neutron Application Reactor) went into its first criticality in 1995 at the Daedeok site. It was a one-of-a-kind 30 MW thermal open pool-type reactor with 19 percent enriched fuel, fully in line with the international nonproliferation guidelines, designed jointly by KAERI and AECL of Canada.

During the 3rd and 4th Republics under President Park Chung-hee in the 1960s and 1970s, high national priority was given to science and technology developments in support of rapid economic growth. The master plan to develop a new science town away from Seoul (thirty square kilometers of government land near Daejeon) was initiated in the late 1960s, followed by the relocation of many national research institutes such as Korea Standards Research Institute in 1974. In the nuclear sector, the spin-off institute KNFDI first moved to Daedeok in 1976 then was followed by the main KAERI relocation beginning in 1981. Today Daedeok Science Town is home to fifty research centers from all science and technology areas, both national and private sectors, with technical/national universities built-up since the 1970s. It was modeled after similar science cities abroad, such as Tsukuba of Japan and Research Triangle of North Carolina, USA, to maximize synergy among national laboratories, academia, and selected industries. Today, high-tech venture companies are moving into the northern suburb of Daejeon (the sixth largest city in Korea, with a population of 1.6 million), earning the nicknames of "Daedeok Valley" and "Daedeok Innopolis." Daedeok Innopolis is now home to twelve national/public nuclear technology/

research entities besides KAERI; the nuclear regulatory agency Korea Institute of Nuclear Safety (KINS), nuclear fuel fabrication company Korea Nuclear Fuel (KNF), power reactor systems design division of Korea Power Engineering Company (KEPCO-E&C), Nuclear Engineering Technology Center of KHNP, research institute for KEPCO Korea Electric Power Research Institute (KEPRI), most recently established national nuclear accountancy control agency Korea Institute of Nuclear Nonproliferation and Control (KINAC), Nuclear & Quantum Engineering Department at KAIST, Tokamak nuclear fusion center K-STAR, National Research Foundation, and the nuclear academic society Korea Nuclear Society (KNS) are all located in the Daedeok Innopolis, Daejeon. In addition, a number of small nuclear venture companies in software developments and hardware service and production are clustered around the Innopolis, known as the "Nuclear Valley." Starting with the reluctant move away from Seoul thirty years ago, today Daejeon is playing the role of technical hub not only to major Korean nuclear programs, gaining the reputation of "Nuclear Technology Mecca" of Korea, but also in other high-tech areas such as information and bio-technologies. (see "Map of Daedeok Innopolis – Nuclear Technology Mecca" in page ii.)

Nuclear Safety Center

The nation's very first NPP Kori-1 was in full operation since 1978 while the regulatory safety infrastructure was still at its infancy under MOST. Foreign advisors were also invited by MOST and provided invaluable service to the Korean nuclear regulatory infrastructure from the early 1970s. IAEA dispatched the first safety expert mission to Korea in 1973 composed of four experts: Morris Rosen of USAEC, Abel Gonzales of CNEA Argentina, Murray Duncan of AECB Canada, and Ishikawa Michio of JAERI, Japan. Their main mission was to provide a wake-up call to Korean authorities on the importance of indigenous safety regulatory infrastructure, starting from reviewing the Kori-1 PSAR dockets. Interests of the NSSS supplier Westinghouse did not always coincide with the Korean side. One notable person was Morris Rosen who came back and worked in Korea for over a year in 1974 as an IAEA expert in nuclear safety. His understanding of the technical issues together with Korean culture and mentality earned him much respect and friendship, contributing greatly to the Korean nuclear safety culture at the early stage. Later he served as the Deputy Director General of Nuclear Safety at the IAEA in the 1990s, a new post created following the Chernobyl accident.[4]

KAERI, being the only nuclear entity besides the nuclear operator KEPCO, was expected to take the lead role in assessing the safety technologies of a giant new nuclear power plant independent from the operator. Trying to understand how the plant was built under which safety criteria and standards was the first order of business. Several sections were newly established at KAERI to study safety aspects of the NPP from the mid-1970s: a thermal-hydraulic laboratory for severe accidents, nuclear fuel design laboratory for in-core management, and applied mechanics laboratory for seismic and structural analysis of major NSSS components. These labs were managed by newly recruited US-educated Korean engineers from well-known universities like MIT, Caltech, RPI and Penn State. The Korean government was eager to coax well-educated, highly experienced Korean nationals from the US in the mid-1970s, when the country was building its first NPP, offering them special privileges like company housing and transportation.

The TMI accident in April of 1979 in Pennsylvania, USA, brought global awareness of nuclear safety. Hundreds of TMI Action Items were introduced and implemented by the USNRC and nuclear industries worldwide, leading to the creation of the Institute of Nuclear Power Operators (INPO) in the US. It was a huge shock to the Korean nuclear sector at that time as well since the accident originated in the US on a PWR-type NPP supplied by Babcock & Wilcox (although not from the same nuclear vendor company Westinghouse for Kori-1). The wake-up call in nuclear safety was loud and clear when the issue of "close down KAERI" caused a desperate search for alternative solutions in 1981. The main mission of KAERI was about to shift from basic science to nuclear safety.

Director General Kang had a special visitor from the US at his MOST office in late 1980, again escorted by Science Officer Stella of the embassy. He was Dr. Davis, a high-ranking USDOE officer, seemingly in line with Kang's insistence on saving the peaceful use argument for KAERI. Dr. Davis made an argument in describing the status of the Korean nuclear regulatory situation as "a blind man standing on a cliff without knowing one step ahead before falling down" to dramatize the lack of technical infrastructure in managing the safety of nuclear power, especially following the TMI accident. Kang and Davis apparently reached an agreement in supporting nuclear safety for Korea on an urgent basis in safety manpower training, establishing a MOST/USNRC hotline in case of accidents, and emergency exercise procedures.

Gaining confidence in winning the case for the survival of KAERI, Kang called KAERI President Cha Jong-hee, informing him of the impending changes on the name change (the merge with KNFDI), the physical relocation to the Daedeok site, and a need to create a new nuclear safety center. At that point Kang could not tell the exact reasons those historic changes had to take place immediately, and no one dared to ask. President Cha had to make a hard bargain with MOST, notably the government commitment, to finance a new research reactor at Daedeok as a condition for the name change and decommissioning of two research reactors (TRIGA Mark II & III) before the move down to Daedeok. A task force team was created in KAERI for a new entity in nuclear safety technologies, leading to the formal opening of the Nuclear Safety Center (NSC) at KAERI in February 1982, with Kim Dong-hoon as its first director. NSC served its role of conducting technical nuclear regulatory review and inspection services to MOST until it became the fully independent Korea Institute of Nuclear Safety (KINS), away from KAERI, in 1990.[5]

All in all, the 1980 crisis of KAERI's very existence, apparently caused by the Chun Doo-hwan's new military government, found a way out of this predicament to put out the immediate fire, thanks primarily to several faithful civil servants like Kang. The three-point strategic alternatives:

the name change from "Atomic Energy" to "Advanced Energy", commitment to relocate to Daedeok from the Taenung campus, coupled with the creation of the new Nuclear Safety Center at KAERI were apparently convincing enough to Minister Lee and the Blue House. Moreover, the alternatives must have given enough confidence to the US that KAERI was indeed restructured to erase any past proliferation doubts of the 1970s once sufficient transparency measures were in place.

History is full of ironies. This dramatic episode of KAERI's closure in 1980 was about to prove itself to be a "blessing in disguise" as the subsequent turn of events would unfold for truly peaceful, commercial Koreanization of nuclear power technology in the next thirty years.

Lingering doubts

Although the KAERI program on the fuel cycle pilot plant and material testing reactor construction was formally ended by 1975, it took several more years to bring alternate programs to full speed for reallocation of budgets and manpower during the late 1970s. Popular sentiment on the development of the Korean nuclear capability was still lingering among some as a "sovereign right" or "self-defense" argument. A best-selling fiction book in Korea during the 1990s was *Mugungwha Flower* (hibiscus, the Korean national flower) *is Blossoming* by Kim Jin-myung, based on a fictitious national nuclear program supposedly led by the much-respected Korean-born elementary particle physicist Dr. Benjamin Whisoh Lee (1935-1977) due to his untimely death by a mysterious traffic accident in Illinois in 1977. The Korean Broadcasting System (KBS) televised special science month documentaries on the "Secrets of Lee Whisoh" in April 2010, after thirty-three years, putting an end to speculations that his death was somehow linked to a conspiracy plot to stop the Korean nuclear program. It was finally proven that it was a simple car accident and nothing more, saving his honor as a world's top scientist, and disconnecting his name from any specific fuel cycle program connection. A national drive is underway to establish the "Lee Whisoh Center for Physics" in Korea for pure research, honoring the late scientist.

The US government has been playing a vigilant role in global nuclear nonproliferation, more so in regions like the Korean Peninsula and, during turbulent times in political changes during the 1975-1995 period. The US Embassy in Seoul played a crucial role in monitoring major nuclear activities in the country. Science attachés (or science officers) at the Embassy were without exception selected with nuclear expertise/backgrounds, from USDOE labs, USNRC, or intelligence agencies. Robert Stella (1976-1980), Robert Liimatainen (1981-1984), Jerome (Sam) Bosken (1984-1988), Kenneth Cohen (1989-1993), and Kenneth Crosher (1993-1996) were the science attachés, all very knowledgeable in nuclear technologies and also well known to the Korean nuclear community. Subsequent US science counselors in the Seoul Embassy changed their expertise/background from nuclear to other technologies since the late 1990s, showing the changing interests of US monitoring. KAERI's management was rather "open door" to all relevant foreign visitors in showing and explaining what was going on and what the future plans were since 1980. This open-door policy was kept to enhance the transparency of nuclear R&D matters, which once was confidential on special projects during the 1970s. The science officers made frequent visits to nuclear sites, including Daedeok, while the new nuclear center was being constructed and new facilities were under commissioning. Being knowledgeable in nuclear matters with welcoming hosts, they were able to assess the very nature of the nuclear R&D program in Korea at that time.

One notable person remembered well by many KAERI staff members was John (Phil) Colton from the US State Department, Arms Control & Disarmaments Agency (ACDA). He stayed at the KAERI Daedeok campus in 1983 on an IAEA expert mission for six weeks as a nuclear fuel QA expert while the newly built heavy water reactor (CANDU-type) fuel fabrication plant was under commissioning stage. He assisted the fuel pilot plant production process into maturity, but also became cognizant of overall KAERI operations as well. He earned confidence and trust among his hosts by the end of his stay at Daedeok, overcoming mutual suspicions at the beginning. He provided not only much-needed fuel fabrication QA expertise to KAERI but also helped to demonstrate transparency to the outside by his positive feedback. His recollection on the 1983 visit sheds some light on his personal observations and what he saw then at KAERI:

> "While I was there, I was happy to note that Dr. Han Pil-soon made the facilities totally open to me. I was not tasked to "inspect" the facilities, but the openness went a long way to eliminate suspicions which might have occurred if parts of the facilities had been closed to me. I noted that you (BK Kim) also gave me the impression of openness while others treated me, initially, with suspicions. They seemed surprised that the director (Dr. Han) would want to be so open. I would have had to be blind not to have noted facilities such as the separation facilities with pulse columns, but that was also a dual-use technology where cold scrap from the pelletizing facility would need to be re-dissolved and recycled back to the process and those could be justified. Once again I stress that I was there as an expert and not an inspector. However, what actions and openness I observed could/would influence the attitudes of those looking at/for other activities later."[6]

Another vigilant visitor to Daedeok in 1985 sent by the US State Department was Professor Alan Sessom (physics PhD from Yale University), a handsome Afro-American physics professor from Harvard University. He was probing the real reasons why KAERI selected KWU over Westinghouse for the PWR technology transfer partner, while Westinghouse made a much favorable offer. Han told Sessom, "Yes, we recognized the exceptional bid from the US with 'A' grade, but Germans did it better with 'A+' grade." Then went on to convincing the visitor the need to get to the technical self-reliance in the fuel technology for a resource poor country like Korea. After several visits to Daedeok, Sessom seemed convinced of the Korean intention, provided the transparency policy is maintained.[7]

More official bilateral meetings were instituted between the ROK and US governments since 1977, the Joint Standing Committee on Nuclear and Other Energy Technologies (JSCNOET), headed by MOST, and the US State Department. Annual meetings held at alternating venues (in Washington DC or Seoul) now matured into a model nuclear bilateral meeting (2009 earmarked the thirty-third JSCNOET meeting). This forum contributed significantly in providing linkage among non-commercial nuclear entities in two countries, ultimately building confidence though transparency policies from both sides in all aspects of nuclear technologies. Many technical experts from USDOE national laboratories and USNRC and their Korean counterparts exchanged technical visits through the JSCNOET platform. One notable person in this bilateral cooperation is Alex Burkart (PhD in nuclear engineering) of US State Department, who served on JSCNOET for over thirty years, from the very beginning as a US representative, providing nuclear expertise as well as continuity. It became clear to everyone's mind that maintaining absolute transparency while building up friendship was the best option to win the confidence of peaceful uses of nuclear research. This laid a solid foundation of nuclear power technical self-reliance drive in the coming decades.

CHAPTER 3: SAVIORS:

Military academics

Any major nuclear country's program is bound to have some military dimensions in many aspects: political, technical, and human connections. Immediately following the Manhattan Project after the Second World War, the big powers embarked on national-scale nuclear development programs for weapons and nuclear propulsion, as well as commercial nuclear power. Nuclear weapons programs are difficult to trace as they are kept in total secrecy. However, for peaceful uses in nuclear power, especially after "Atoms for Peace" in 1953 and the creation of the IAEA in 1957, they are coming out in an open and transparent world for further cooperation and developments.

The most notable person who pioneered the early power reactor program in the US was Admiral Hyman Rickover (1900-1986, US Naval Academy class of 1922) from the US Navy. He left a legendary name for developing nuclear propulsion for the nuclear submarine fleet, but more so for creating the culture of the highest quality and integrity in nuclear power industries for the coming generations. His submarine propulsion system was the forerunner of the modern pressurized water reactor (PWR) system, further developed and commercialized by Westinghouse. PWR-type nuclear power plants are the world's dominating NPP types today (about 70 percent of all operating NPPs worldwide). The same reactor technology was licensed by Westinghouse to Framatome of France and Mitsubishi of Japan. Many veterans of the US nuclear navy served the US nuclear power industries as living guardians of quality assurance and integrity, which have become synonymous with nuclear power industries. Vice Admiral Castro Madero of Argentina was also the founding figure for the Argentine nuclear program, serving as the head of the Argentine National Atomic Energy Commission (1976-1983), developing indigenous nuclear fuel and power programs in the 1970s. These are some examples of how military academics (mostly from the navy) have contributed in the dawn of nuclear energy worldwide. In fact, most military academic institutions teach, in their regular curricula on "nuclear strategy," courses to educate their future military officers, covering the basics of nuclear weaponry, radiation hazards, nuclear energy for electricity production, etc.

The nuclear power program in Korea had its beginning in the late 1960s with the construction of Kori-1 NPP during the administration of President Park Chung-hee (graduate of Japanese Military Academy in Manchuria in 1942, when Korea was under Japanese colonial era). President Park is remembered as the political leader who brought Korea out of hunger and poverty, and laid a foundation for the export-oriented industrialization from an agriculture country. He also recognized the importance of science, technology, and energy independence including nuclear electricity production since the first oil crisis of 1973. His assassination in 1979 led to yet another military government by President Chun Doo-hwan for the coming decade. Despite its political setbacks in democracy, however, history has it that they laid a solid foundation for a world-level thriving economy (joining OECD in 1995, G20 in 2009) together with civilian democracy. In hindsight, during the 1980s under Chun's government, the Korean national economy experienced record stable growth, together with a large expansion of its nuclear power program based on self-reliant indigenous technologies. Several key figures during this period played crucial roles in shaping the nuclear policies from a foreign technology base to domestic infrastructure-building since 1980. They are the foursome starting with President Chun Doo-hwan (1980-1988), minister of MOST Kim Sung-jin (1985-1986), President of KEPCO Park Jung-ki (1983-1987, nicknamed "Rocky Park"), and President of KAERI Han Pil-soon (1984-1991) in their respective capacities in office. This Chun/Kim/Park/Han foursome made unique and personal contributions in the 1980s to lay foundations to make the "Koreanization" of nuclear power technology possible. They all happened to come out of military academies in Korea. Chun and Kim were classmates from the same eleventh class of the Korea Military Academy in 1955, Park was from the fourteenth class of KMA in 1958, while Han graduated from the Air Force Academy in 1957. During the 1950s and 1960s, when Korea was still at an economic poverty level following the devastating Korean War, many bright high school graduates chose military academies for university education instead of more prestigious ones like Seoul National University. Many military academy graduates left their mark in science and technology administration with high distinction during the unprecedented rapid economic developments since the 1970s.

This book will unfold the saga of how nuclear power technology was saved from politics in turbulent times in a developing country like Korea, guided by Chun and Kim in the background, and orchestrated by Han and Park for nuclear power technology developments as the heads of the national nuclear research center and the sole nuclear electric utility, respectively, in the 1980s.

Although the top diplomatic records in the early days of Chun's administration is still not declassified yet, the following assumptions can be made. It had to be so urgent to win US recognition of the new regime in 1980 to force the closure measures at KAERI, which resulted in removing the word "Atomic" from the official title of the institute, among other things. It appeared a bitter pill to swallow for Chun and a huge setback for KAERI at the time. However, it turned out to be quite the reverse in the following decade. President Chun must have felt personally sorry to the people at KAERI for what he had to do. His policy actions in nuclear power technology developments provide ample evidence in testifying to his commitment to self-reliant technology build-up by appointing the right persons to key posts, like Han and Park with full authority. One may say Chun paid back handsomely after his initial mishaps in the "Atomic" connection with unprecedented policy support, which led to the Korean success story. Another crucial person who played the advisor's role to President Chun on nuclear matters was Kim Sung-jin, as Chun had absolute trust and confidence in Kim's policy recommendations. Kim was a legendary super cadet of KMA, known as "Gyobeon Number One," not only being the top in academic ranking in KMA history, but also a most respected strategist.

Kim was well known among all KMA graduates not only for his academic score, but also as the most respected cadet due to his character and leadership. When the May 16 Military Revolution broke out in 1961 under General Park Chung-hee, Kim was serving as a professor at KMA, opposed to the revolution in principle in that KMA should not be involved with politics. Due to this episode, he was sidelined by the new military regime, missing a promotion to a general's rank. Instead, he was sent to study abroad, and pursued his MS and PhD in physics and mechanical engineering at University of Illinois and University of Florida, respectively. He was the first case of a Korean military officer sent abroad for his doctorate. He met Han Pil-soon as a fellow classmate at Illinois while studying physics together, both with ROK government scholarships to military officers. They shared the same university dormitory, staying up late at night discussing homework assignments, often times arguing with each other. Their friendship and confidence in each other played an important role during the dynamic years of the Korean nuclear program in the 1980s.[1]

Kim was the point man for Chun in Washington in winning the US diplomatic recognition in 1980, and later briefly served as the minister of MOST from 1985-1986. Kim is remembered as a special mentor to Chun, Park, and Han in providing crucial advice behind the scenes in the areas of science and technology in general, particularly in nuclear policy decision-making, with the highest confidence from the top. With his stewardship among the key players in the most transforming period of the 1980s, the subsequent nuclear power technology development with the brain power of KAERI, combined with financial and project management power of KEPCO was the only possibility. It is most unfortunate that this book could not benefit from a personal interview with Kim due to his untimely death in 2006.

All in all, the Chun/Kim/Park/Han quartet made part of a unique history unparalleled in any developing country during the 1980s for achieving the dream of technical independence in nuclear power, overcoming multiple shortages of experienced manpower, time, and money. This book will show how Chun and Kim provided the policy support behind the scene, while Han and Park managed the project goal deliveries, as well as technology self-reliance, as two field marshals.

KAERI's new boss

Han Pil-soon was born in 1933 in Gangseo, North Korea, and took refuge to the south during the Korean War in his late teens. After graduating from the Korea Air Force Academy in 1957, he went on to complete his graduate studies in physics from the University of Illinois (MS) and University of California, Riverside (PhD) under an ROK government scholarship. Many bright young students chose military academies in the 1950s as most conventional universities could not function with normal academic programs during the Korean War period. His strong academic excellence convinced his superiors in the Air Force Academy to grant him a chance to pursue higher education instead of a real military career. Upon returning back from the US, he served at the Agency for Defense Development (ADD) under the Ministry of Defense in Daejeon for twelve years. He was one of the best known for the Koreanization of many military hardware like hand grenades, bullet-proof helmets, and fixing malfunctioning high-tech weaponry like the Vulcan cannon in the 1970s. His talent in new design formulation to suit the demanding Korean field situation, then his ability to complete the missions in full field deployment, was well demonstrated. He knew how to motivate the people around him to outperform themselves and gain full trust. He was one of the star scientists recognized by the MOD and the Blue House under President Park Chung-hee's self-reliant national defense, moving up to the rank of Air Force colonel and vice president of ADD in his forties.

Han's appointment as the vice president of KAERI in charge of the Daedeok Engineering Center (the old KNFDI at the Daedeok site) in March, 1982 came as a surprise even to him (he was forty-nine) while ADD was undergoing massive downsizing layoffs following the new military government take over in 1980. He welcomed the new opportunity to serve in the nuclear energy research center, being a theoretical physicist inspired by quantum mechanics and nuclear physics. His arrival at KAERI came at the lowest morale point since the restructuring of KAERI under the new President Chun's government, which came only a year after the name "Atomic" disappeared from KAERI's name. MOST was cutting back the research budget with unsympathetic eyes due to lingering suspicions of the plutonium program of the past, and was not able to support the newly emerging nuclear power utility in a significant way. Many able scientists were voluntarily resigning from KAERI for better opportunities at universities. DEC campus was nearly deserted after 5 p.m., showing definite signs of a declining research center. In early 1982, government plans to relocate the old KAERI Taenung campus were made firm, and the KAERI's center of gravity was shifting from Seoul to Daedeok, although the head of DEC was reporting to the KAERI president in Seoul then. Only a couple of research sections voluntarily moved down to Daedeok while the majority of the KAERI Seoul staff showed reluctance to relocate away from Seoul. It was understandable that most research and city infrastructure was lacking at the Daedeok Science Town in the early 1980s.

Han soon discovered the overwhelming negative "cannot do" spirit prevailing among KAERI staff at that time. This is when his unique management style started to make the difference in motivating his staff by turning the "cannot do spirit" into "can do spirit" with his step-by-step approach. His first task was to restructure the existing research projects, overall about five million dollars' worth scattered among forty something different projects (KAERI DEC had a total working staff of about three hundred in 1982). One may recall the "Wolsong Fuel" Project, which got off to a good start in 1980 during the KNFDI era, just before Han's arrival. Its mission was to produce prototype fuel bundles in two years. Still, most project team members had no vision of the future beyond the demo production for real loading into Wolsong NPP. Han managed to motivate the team by revising the two-year plan to five years, to the end target of Wolsong NPP loading and mass production for full core (Wolsong Unit 1 was under construction, scheduled to go to commercial operation by 1983). By mid-1982, more than half of the total budget and manpower resources of KAERI-DEC was integrated into one single "Wolsong Fuel" Project, with one hundred and fifty full-time staff members. The project team was organized as below, with Han himself taking the overall project manager's role, and team leaders at the beginning. They were all in their thirties and forties (in parentheses, their current affiliations are shown):

> Fuel design: Suk Ho-chun (now with Canadian Nuclear Regulatory)
> Fuel fabrication: Suh Kyung-soo (after the project completion, died of cancer in 1988)
> In-core management: Kim Seong-yun (served as KAERI president in 1996-1999)
> Out-of-pile test: Kim Byung-koo (author of this book, KAERI VP in the 1990's, director of IAEA in 2002-2008)
> Uranium conversion: Chang In-soon (served as KAERI president from 1999-2005)
> Quality control: Lee Gyu-am (later served as VP of Gaia with Han Pil-soon)
> Project coordination: Nam Jang-soo (now director general of Korea Nuclear Society)

Never before had KAERI witnessed so many dedicated staff fully motivated for a single mission: to develop and deliver nuclear fuel usable in Wolsong NPP. Note that all team leaders served in much higher capacity after the Wolsong Project completion. It was a "reverse engineering" project in reality, since to develop local fuel-supplying capability with Canadian assistance would have come with a high price tag (AECL has quoted Canadian twenty-five million dollars for technology transfer). This propelled even stronger determination to "we can do it" for obvious reasons: to save scarce foreign currency as well as to demonstrate that "we can do." Prototype fuel assemblies were fabricated with natural uranium pellets and zirconium tube parts. The initial batch of yellowcake was processed at the KAERI site under Chang In-soon's conversion team. They built the chemical processing line from the published academic papers, but no detail engineering data was available in open literatures, but determined to succeed in commissioning a pilot scale conversion plant to produce the uranium oxide powder (needed for the fuel fabrication process) from yellowcake powder. With repeated trial and errors, Chang's team developed indigenous conversion process technology which became the backbone of the commercial scale conversion plants at KNFC in the 1990s. Chang recollects one special trip with a well-known German conversion specialist, Dr. Helmut Assman, to Gyeongju in acquiring crucial technical information. Upon Chang's genuine inquisitiveness by invoking his personal pride of a scientist, Assman finally gave out the information with the written promise to use it only for the KAERI project.[2]

One other Koreanization episode on the Wolsong fuel project in 1981 needs to be told here. The KAERI project team at that time was in a desperate need of full access to the real CANDU fuel and fuel channel which only existed at the Wolsong construction site (it was just prior to the initial core loading) in order to verify the geometric/mechanical integrity of their prototype fuel. They made a decision to plea to the KEPCO site office, fully aware of the difficulties of accessing the nuclear fuel which was still under the AECL control, as it was a full turn-key project. Special KAERI project team (including myself) with metrology instruments went to the Wolsong site and met the KEPCO site superintendant Suh Suk-chun in private, and explained the need of nuclear fuel measurements for the localization purpose. Suh, an electrical engineer from SNU (later served as the President of KPS, the NPP maintenance company), listened to the plea, thought about a while, then told us to come back at midnight, suggesting a clandestine operation at the fuel storage vault. He must have thought the risk of breaking in the vault was worth taking, to help out the localization team to verify the locally made CANDU fuel against the original. Only the nondestructive metrology measurements were taken as any destructive testing could invoke the IAEA safeguards accountancy control. It was in the critical path for the localization team to move on to the next step of the out-of-pile verification testing. The Hot Test Loop to simulate the Wolsong flow conditions was being built at Daedeok with the full size Wolsong fuel channel, with its key component pressure tube supplied from the KEPCO's operational spare. Suh's bold decision to help out the fuel localization effort was the triggering element of a long march to the NPP technical self-reliance as we know it today. In a recent interview, I asked Suh what was his motivation to help out the localization team, clearly risking his own job. Suh said, "I knew it was right thing to help out the KAERI team, although it may be against the contract requirements. On that evening I quietly called Jerry Sovka, the AECL Engineering Manager at the site responsible for the fuel custody, explained the need to examine the fuel. Sovka, an immigrant engineer from Czech himself, somehow understood the localization need of a developing country, agreed to open the box, provided the missing bundles be brought back in time". Suh continued, "I had to notify the responsible officer to be clear in my conscience from 'stealing' anything". It was a remarkable testimony from an old engineer brought to light for the first time after thirty years. Like-minded engineers from KEPCO, KAERI and AECL did come together in this one episode without fully knowing its far reaching implications for decades to come.[3]

Initial formal response from KEPCO head office was rather negative, as expected from any conservative utility company. Simply put, "You must be joking. How can we trust a locally produced, unproven nuclear fuel in risking the Wolsong performance warranty?" It went further: "You must produce two ministers' (MER and MOST) guarantee signatures before we can let you burn the local fuel." The credibility gap between

KEPCO and KAERI was huge, to say the least. MOST Minister Lee Jeong-o played a crucial negotiation role in convincing KEPCO to take the locally produced fuel for Wolsong loading, provided KAERI would pass the verification testing requirements by AECL in addition to the regulatory approval on the local fuel. Out-of-pile testing was started at the newly-built Daedeok Hot Test Loop for the mechanical integrity of the fuel under the reactor flow conditions. Serious negotiations started with AECL for the in-pile testing service at their Chalk River NRU reactor. Han was determined to succeed in his first mission.

The Wolsong fuel test bundles began to turn out at the fuel fabrication pilot plant at Daedeok around January 1982. They passed all physical, material specifications in ambient conditions, as any good reverse engineering prototypes can hope to achieve. The next goal was to demonstrate its durability under the reactor coolant conditions: high temperature, high pressure, and flow rate as in the Wolsong primary loop. Initial contact with AECL quoted nearly one million Canadian dollars for the out-of-pile testing at AECL's Sheridan Park Engineering Laboratory. For that kind of money, we could build our own hot test loop, they thought. An out-of-pile testing Hot Test Loop facility was built using an abandoned motor pool building at Daedeok in less than a year's time. A full size CANDU fuel channel was installed for the loop simulation, thanks to KEPCO and Hyundai Wolsong construction team, to provide the pressure tube and the end fittings. After the two-thousand hour endurance test examination, the test bundles survived the harsh conditions. They passed the second test requirement at home, with the indigenous facility and manpower.

The next and final step was to conduct an in-pile testing where actual neutron fission reaction was to take place. They had to prove the test bundles would also reach full burn-up and still maintain physical integrity. The only place where CANDU fuel bundles could be in-pile tested was the NRU reactor fuel test loop at Chalk River, Canada. The initial quotation from AECL was Canadian three million dollars, which was about two years' total research budget at KAERI at that time. It was clearly out of reach. After an episode by Han and Robert Hart (VP of AECL) at the Montreal Airport for price negotiation, an in-pile test service agreement was reached at four hundred thousand Canadian dollars. It was the ultimatum price for which Han was authorized from Minister Lee. After seven months of in-pile testing at NRU, the KAERI fuel bundles were declared "fit for reactor service," meaning they passed all performance irradiation testing. They were the final provisions to convince KEPCO of the "reactor worthiness" of the local nuclear fuel. Finally KAERI fuel bundles were loaded into Wolsong NPP on September 1984, followed by full burn-up discharge one year later. It was a defining, jubilant moment for all the Wolsong Project team.

President's visit

One historic event took place at Daedeok on April 12, 1983 which left a lasting impact. President Chun Doo-hwan paid a personal visit to KAERI's pilot fuel fabrication plant, under short notice. Han briefed Chun on the Wolsong Fuel Project progress to the real Wolsong loading shortly, quoting my phone call from Chalk River the day before that the test bundle irradiation was a proven success. After taking a technical tour of the production line, Chun asked Han, "We have more PWR plants than PHWR. Then why are you only making CANDU fuels here?" Han replied, "PWR fuels are more difficult and are to be made at KNFC, a newly formed subsidiary of KEPCO, to bring in foreign technology, sir." Chun appeared a bit puzzled, then asked, "Tell me why Koreans cannot develop PWR fuel if you can do the CANDU fuel, KAERI's approach may bring better self-reliance". Then he left Daedeok.

Impact of this short visit of 1983 turned out to be bigger than anyone expected. In retrospect, it was a pivotal turning point for KAERI to take up the commercial NPP projects solely for peaceful purposes. Han was asked to present the overall nuclear technology status to the Blue House (the presidential palace in Seoul). Two months later, Han received a phone call from the Blue House informing him to be the next president of KNFC. He was appointed as the second CEO of

KNFC and the KAERI president at the same time. It was the real recognition of the Wolsong Fuel Project's success under Han's leadership and a strong endorsement for his new approach to PWR fuel localization based on Korean technical self-reliance. President Chun appeared to have shown his determination for nuclear technology self-reliance by giving a chance to an indigenous technology zealot like Han. I had a strong impression that it might have been his way of redeeming his feelings of guilt over his order to close down KAERI back in 1980. What started out to be a serious blow that almost killed KAERI as the national nuclear laboratory, for the sake of saving the legitimacy of the new born regime, was about to turn the other way around for the indigenous technology and the nation's nuclear future, I thought. The president's overt showing of support at this point must have had a magical power in reshaping the future of Korean nuclear technology from that point on.[4]

First nuclear hero

One person who stood the tallest among the initial Wolsong fuel team was Suh Kyung-soo (1940-1988), the section head of the fuel fabrication laboratory. He graduated from Korea University in Seoul with a PhD in physics, and joined KAERI in 1970 in the fuel metallurgy laboratory. He was among the first to volunteer and move down to Daedeok when the first Wolsong fuel localization project got started. He had dedication and enthusiasm to prove the reverse engineered prototype bundle would be high enough quality in reactor worthiness. He worked day and night with his team at the fuel fabrication pilot plant that was under construction, with the help of the French company CERCA. He was the type of person – a "benevolent workaholic" to set the example for others to follow. When the first test bundles with natural uranium were produced by his team, he said, "This is like my offspring. I want to see through to the end of this first local fuel in Wolsong." When the opportunity came to ship the test bundles to AECL for the irradiation testing, Suh and I were selected to hand-carry three bundles in person to Chalk River. Suh made a comment before our departure from Daejeon, saying, "I will jump into the Pacific Ocean and will not come back home if my fuel bundles do not pass the irradiation testing." It was in late March of 1983. We took transfer flights through New York and Toronto, then finally arrived at a tiny municipal airport at Pembroke near Chalk River, Canada, a remote nuclear research laboratory at the northern end of Ontario Province. After the initial qualification testing, two test bundles (one natural uranium, one enriched uranium of 2.5 percent) were loaded into the U2 loop at the NRU reactor on the morning of April 9, 1983. It was a moment of pray for both of us to witness, from the top of the reactor pool, the fuels being put into the test channel. A Chalk River fuel expert already told us any probability of irradiation failure would occur within three days of the initial loading. Any sign of above-normal levels of radiation monitoring from the U2 channel would indicate such failure.

We felt the following three days seemed like an eternity. Finally, the reactor operator called us at the AECL guest house, informing us that no radiation leakage was detected on the third day. Now our babies were in for long-term burn-up testing. We were jubilant, embracing each other, thanking our AECL hosts. We wasted no time in placing an overseas long-distance call to President Han at his home at 3 am Korean local time to inform him of the good news. As it turned out, Han quoted our phone message to President Chun Doo-hwan the very next day when Chun paid a historic visit to the KAERI site for the first time, giving President Chun the assurance of indigenous nuclear technologies. The good news from Chalk River prompted Han to make his next steps in the Wolsong fuel mass production, LWR fuel localization project, and ultimately winning the chance to conduct the reactor systems design work at KAERI.

Upon our return from Chalk River in mid-April 1983, the Wolsong fuel project moved into the next stage: the mass production to scale up the ten ton-per-year capacity pilot fabrication plant to one hundred ton capacity to supply the full core of Wolsong Unit 1 from 1986. By this time KEPCO was pushed to accept the locally fabricated nuclear fuel, provided the Chalk River irradiation tests turned out satisfactory and the regulatory licensing approved the compatibility of the new fuel. Suh was the natural person to take this next challenge, and he was more than ready and happy to be the project manager for the mass production. New equipment was ordered, staff recruited, and most of all, Suh was preaching the gospel of nuclear quality assurance to be built into the new fabrication process. He was exerting everything he had in ensuring the mass production line to be ready for the one hundred ton production by 1987, without even knowing the malignant tumor was spreading in his body. Much to everyone's shock and sadness, Suh passed away on a clear autumn day in 1988, when Korea was in the midst of its first time ever hosting the Seoul Olympics. He was only forty-eight. On October 9, 1988, his funeral service was held at the KAERI campus amongst sobbing family and colleagues who worked with him for so long to make fuel fabrication history. No other person had the honor to be given an institute funeral other than the sitting president of KAERI. At his first anniversary day, a bronze memorial bust statue was dedicated at Enertopia Park inside the KAERI campus. During this statue dedication ceremony, Han expressed his feelings: "Let us remember Suh as our first and true hero of the Korean nuclear power technology, and more heroes will be forthcoming." Suh's highest position he served was a division director, but he commands more respect and recognition than any other higher-ranking managers at KAERI. Today his statue is still the sole monument honoring a nuclear scientist or an engineer anywhere in Korea.

Erlangen story

Han and his brain powers went immediately to revisit and revise the existing PWR fuel localization plan in the newly formed KNFC. Three basic policy directions were revised:

1. From KNFC being a hardware fabrication company only with foreign design, to bringing the design and engineering work to Korean scope (to be done by KAERI)

2. From the KNFC plan to solicit a joint venture company with foreign partner's 50 percent capital and performance warranty, to a technology transfer via a competitive tender process using 100 percent Korean capital
3. From using foreign consultants to evaluate the technology transfer international bidding process, to an indigenous bid evaluation process.

Technical self-reliance was making a higher profile at the cost of taking much higher risks of failure in fuel performance and warranty. The founding president of KNFC, Kim Sun-chang, who had to abdicate his post to Han at that time, recently recollected his feelings about the technical self-reliance: "In retrospect, Han's timeless efforts in pushing for the indigenous technology approach with more risk-taking did pay off, as he worked so hard to convince everyone."[5]

Han set out to make his second project a success in PWR fuel localization, only one and a half years after he joined KAERI in 1982. A special task force team was mobilized at KAERI and KNFC for preparing invitation-to-bid documents for PWR fuel technology. Vice President of KAERI, Rim Chang-saeng, graduate of MIT in nuclear engineering and experienced nuclear fuel expert from Westinghouse, was the lead figure at KAERI (Rim later served as president of KAERI). About this time, one crucial member joined KAERI with nuclear fuel engineering experience from the US. It was Kim Si-hwan (SH Kim), graduate of SNU and Rensselaer Polytechnic Institute (RPI) in nuclear engineering, who worked at the nuclear fuel division in Combustion Engineering. Immediately upon his return to Korea in 1984, he was given the project manager's role in the newly formed PWR fuel localization project at KAERI. Soon more than a dozen Korean nuclear fuel experts repatriated to KAERI from the US, with friendly persuasion by Kim. He made a unique contribution as the KAERI PWR fuel project manager to the end by recruiting and motivating the best talents in nuclear engineering to initiate the project-oriented divisions at KAERI. Han made a strategic decision to create two separate task forces in selecting the technology transfer partner: one was the Criteria Committee (Rim Chang-saeng of KAERI, Yang Chang-kook of KEPCO, Nam Jang-soo of KAERI), and the other was the Evaluation Committee (Kim Si-hwan, Suh Kyung-soo and Kim Hwa-sup, all from KAERI). The Criteria Committee's role was to come up with the evaluation criteria for the bidding, with their weighing factors (the final signed criteria and weighing factors sheet was kept in a locked safe in Han's office), and the Evaluation Committee's role was to give objective scores on each bidder's proposal based on one hundred points for each criterion. Three evaluation criteria were selected to be: technology transfer conditions (with the highest weighing factor), quality of the technology, and the price. It was clever on Han's part to separate the whole evaluation process into two committees - one to give only the criteria and weighing factors without getting into the evaluation process itself, and the other to process the evaluation without knowing relative weighing factors. ITB documents were sent out to eight well-known international PWR fuel supplier companies, but only five of them submitted the bid based on the evaluation criteria from the ITB document: Westinghouse of the USA, Fragema of France, KWU of Germany, ABB of Sweden, and BNFL of the UK.

After six months of highly competitive bidding and evaluation, KWU was selected as the successful bidder based on their technology transfer and joint design contents. Both committees had done their job successfully. A new concept of *joint design* was first introduced to overcome the shortage of well-trained manpower, budget, and time. KWU was a surprise choice, as it was the least known nuclear company at that time, beating out Westinghouse, which was the dominant nuclear vendor in Korea.

The core of the most elite engineers was sent to Erlangen, Germany, in 1986, to participate in the joint design (while on-the-job training) for all PWR reload core fuel designs in Korea. This was the shining example of how inexperienced but well motivated and educated Korean engineers outperformed KWU expectations in learning and delivering through the joint design process. All in all, by 1989, the KAERI and KNFC alliance was delivering reload fuel design and fabrication service to all PWR plants in Korea. It was the first time in Korean nuclear history that a domestic entity managed the international tendering process from ITB to bid evaluation; the same team at KAERI was responsible for delivering fuel design, and no foreign capital joint venture was needed as one-time technology transfer was sufficient to stand alone.[6]

The second project for Han was turning out to be another success story: being the CEO for both organizations KAERI and KNFC. But this was only a prelude to even bigger technology self-reliance yet to come for the entire NPP system design technology. It was set to be Han's most ambitious project during his nine-year tenure as the president of KAERI (1982 -1991, first two years as VP in charge of DEC). Han is the longest-serving president in KAERI history, laying solid foundations for the Koreanization of nuclear power technology by introducing the system design mentality from nuclear fuel design to NPP reactor system design.

Step-by-step

In his inauguration speech as the KAERI president on April 9, 1984 after two years as VP, Han made his vision for the future and his management philosophy loud and clear. He envisioned the KAERI, sole nuclear national lab, taking an active role in commercial NPP construction projects as fuel designer and NSSS system designer. As a developing country, the only way for Korea to catch up with advanced nuclear nations in technical self-reliance was to engage the best-educated nuclear manpower (they were mostly at KAERI in 1980), with real project experience in design and construction of NPPs. His well-known "hitchhike" strategy, to import the best proven commercial technology from abroad, was to learn the "know-what's" and "know-how's" in copying technology, then move on to creative research and engineering for gaining "know-why's."[7]

Step-by-step, one success after another, it was becoming increasingly clear that Han was no ordinary president of KAERI. He was able to convince MOST and KEPCO to win the crucial decisions on behalf of technical self-reliance. Morale at KAERI was shooting up high, bringing back the "can do spirit" to dream of even bigger success. Many new, talented engineers were recruited to join the growing momentum. Blessings from the very top were filtering down in many subtle ways, from the initial appointment of Han to KAERI and KNFC, then another localization champion to KEPCO, and more to come.

Leadership is about human relationships, one leading many others for a noble cause. One aspect of Han's leadership that captivated his followers was his personal concern for the staff's wellbeing. Han's unique solutions to housing problems could illustrate this point. When the Daedeok Science Town was built in the mid-1970s, nearly everyone stayed at the dormitories or company apartments newly built near research centers. Most of the male staff had their families in Seoul, staying in Daedeok for weekdays, then going back home for the weekends. Public transport to Seoul was always jam-packed on Friday evenings and Monday mornings, full of tired 'Daedeok bachelors'. At the beginning of the 1980s, when KAERI was starting to relocate, existing company housing was already crowded; supply could not meet the demand. Typically, company housing was provided for staff with rental subsidy. Most institute CEOs relied on dormitories and company apartments while their staff had to endure great loss of time and energy to commute back and forth to Seoul every weekend. Working at Daedeok research centers was considered temporary due to complete lack of private housings and living infrastructures available. Then came the golden opportunity to acquire 'my home'. The Korean government developed a huge lot in Doryong-dong, a central location in Daedeok Science Town, for the research staff's family residential zone to be sold to individuals. Although housing needs were acute, no one applied for the land acquisition for their private homes for simple reasons: very few had enough money to buy in lump sum (as long-term mortgage loan system did not exist then), and social infrastructures in schools, hospitals, and general living conditions were primitive compared to Seoul. As a result, no one dared to sell their private properties in Seoul to buy a lot in Daedeok and build their own house. It was considered a poor investment as long-term prospects of living in the Daejeon area did not appear promising.

Han's leadership began to show in this housing issue soon after he was appointed to KAERI in 1982. He set an example of living in Daejeon himself with his family first (only few other CEOs had their family in Daejeon; all were commuting to/from Seoul then). He had the unique idea of a "ten-minute principle" derived from the Niels Bohr Institute, named after the famous Danish physicist, and his fellow scientists living in a community outside of Copenhagen during the early twentieth century. Han's dream was to build a research community where every scientist could spend maximum time in research, being able to commute to his office within ten minutes by car or bus. Doryong-dong residential lot was the ideal location to make Han's dream come true. The only problem was how to mobilize capital funds and convince his staff to put long-term investment value in their own housing in Daedeok. He began to convince Minister Lee Jeong-o of MOST to make interest-free, long-term loans available, arguing stable and permanent living conditions were as important as research itself for scientists. Minister Lee was convinced of Han's argument on the interest free-loans to the scientists, but he needed special blessings to acquire additional budget for the loan from his superiors, notably from the Blue House under President Chun Doohwan. The story of Daedeok housing problem reached the senior Presidential Secretary of Economics, Kim Jae-ik at the Blue House who had all the President Chun's ears at that time. Stanford-educated with PhD in economics, Kim was widely regarded as one of the most important persons contributed to the economic development planning, with the nickname "economic President" under President Chun. Kim saw immediately the value of supporting the scientists to settle down at Daedeok with the home loan package and somehow managed to provide the necessary funding. (This was done only a few months before the most terrible North Korean terror attack occurred in Rangoon. Kim Jae-ik was among the 17 victims who lost their lives at the Aung San mausoleum bombing incident during the President Chun's official visit to Burma in 1983.) The special interest-free loan was announced in 1983, however, despite the terrible loss of the most able civil servants. Han was among the first to buy a lot for himself to set the example. Pretty soon word got around, and KAERI and other institute staff members were taking the loan to buy the residential lots in about ten-minute distance from the office at Doryong-dong. Consequently well over three hundred research scientists managed to take the loan and built their own houses in the 1980s. It was the first successful move of professional scientists and engineers in large numbers away from Seoul in permanent basis with their families. What seemed an impossible dream became reality in less than ten years time, mainly due to one man's vision and courage for the future. Today, Doryong-dong is one of the best known residential areas in Daejeon city, providing a backbone bedroom community to Daedeok Innopolis. This was one of many examples of how the nuclear sector did enjoy the special privileges from the very top of the government under Han's leadership in the 1980s.[8]

Nakamura Sensei

One Japanese nuclear leader, perhaps not well known to the Korean nuclear scene, yet who left a subtle but significant impact on Korean nuclear power development, was Yasuji Nakamura (中村康治, nickname "Koji", 1920-1998). It was through his close personal friendship and collaboration with Han Pil-soon during the 1980s that Han always called him "Nakamura Sensei" ("Sensei" is a Japanese term for "respected mentor," or to end one's name with when you show respect). Nakamura was born in Ishikawa Prefecture, and graduated from the Imperial University of Tokyo in metallurgy when Japan was at its peak of the Second World War with the US. He devoted his entire professional life to nuclear fuel cycle development, plutonium technology in particular, moving up to the chief executive officer of the Tokai Reprocessing Plant at the Power Reactor & Nuclear Fuel Cycle Corporation (PNC). He was among the first Japanese scientists who were trained at the US Hanford Nuclear Laboratory in plutonium technology, where the original weapons-grade plutonium was produced for the Manhattan Project. Nakamura left a major contribution in the plutonium technology development in Japan for their nuclear fuel recycle and fast-breeder development. But perhaps more significant was his crucial technical advisor's role to the Japanese government in reaching the US-Japan Nuclear Cooperation Agreement in the 1970s, allowing the most sensitive fuel cycle technologies, namely reprocessing and enrichment. This is known as the "package prior consent on the use right" of US-origin nuclear materials and technology in the bilateral nuclear diplomacy. Japan was one of the few who had earned this strategic right to use its nuclear materials for the sensitive technologies. It had been one of the top priorities of the Japanese government since the end of the Second World War to negotiate with Washington. Being the most recognized plutonium scientist, Nakamura was behind the several-decades-long diplomatic negotiations, gaining trust and confidence among the politicians and technocrats from both sides of Tokyo and Washington. When he finally retired from PNC in 1982 (he was sixty-two), he was ready to start a new life as he became a technical advisor to Kobe Steel.

Nakamura's first encounter with Korea was through Lee Byung-whie, the late Korean Atomic Energy Commissioner, who introduced him to Han Pil-soon in 1984. The first meeting with him was somewhat odd, knowing his background in plutonium. In fact, during those days, technical contacts with Japan were mainly limited to JAERI in nuclear safety and basic research areas only. Both sides stayed away from sensitive subjects like plutonium recycling and fast breeder technology, thus any contacts with PNC were news items. Nakamura changed that perception from both sides. His recent publication in Japan, *The Choice of Japan: 1984*, was presented to Han at their first meeting. It triggered an avalanche of discussions between Han and Nakamura first, then extended to group discussions with KAERI senior managers. Nakamura's argument on the energy independence through nuclear energy with plutonium recycle was an eye-opener for most Koreans at that time. He gave the example of Japan and how they were able to convince the US to join this fuel recycle projects with absolute awareness and guarantee of nonproliferation. *The Choice of Japan* (in Japanese) was translated into Korean (with a new title, *Conditions for an Energy*

Independent Nation – 1984 Choice of Japan) and published in no time to make it available to wide Korean readership, including politicians and technocrats. This was the beginning of a meaningful relationship with Nakamura Sensei and Korea. He made frequent visits to Korea, mainly to Daedeok and Korean nuclear sites, and also invited Korean nuclear staff to Japan's previously unvisited PNC sites at Tokai and others through his special arrangements.

An exchange of visits to and from Japan and Korea was most active for about ten years for nuclear specialists from both sides. Nakamura organized a number of Japanese technical group visits to Korean nuclear sites under Han's invitation, and vice versa. I was one of the most frequent hosts for Nakamura and his visitors to Korea, and then visited Japanese nuclear facilities under his arrangements. His genuine interest in the ancient history of Korea and Japan during the Baekje ("Kudara" in Japanese) and Aska periods (fourth to seventh centuries) was beyond average amateur-historian level. He had a special notion that the ancient Korean immigrants to Japan were very much responsible for the creation of the ancient Japanese kingdom. I was also much interested in the ancient history between the peninsula and the island, and volunteered to be a local guide to take him to the ancient sites of Gongju and Buyeo (both capital cities of the Baekje kingdom near Daejeon) and Gyeongju (the capital city of Shilla kingdom in the Gyeongsang Province). We were walking on a trail in the Buso-san castle ruins in Buyeo one Sunday afternoon, when I asked Sensei, "Why do these Baekje ruins mean so much to you?" as there wasn't much to be seen for tourists in the 1980s. His reply was simple:

"Baekje is not to be seen with your naked eyes, as they were totally destroyed in the seventh century, but it has to be felt in your heart." Sensei made me feel ashamed to watch him start learning Hangul (the Korean alphabet) and the Korean language when he was sixty-five; (his command of Korean was better than my ability to speak Japanese, though he was old enough to be my father).

When he died of cancer in 1998, everyone who knew him in Korea felt we lost a true friend and a *Sensei* from Japan with much love and respect. He was genuinely interested in helping Koreans as an old friend (rather than a competitor, as a normal situation would dictate), as he must have seen definite possibilities in the 1980s when the Korean nuclear power technology self-reliance drive was just at the beginning.

Han Pil-soon remembered Nakamura Sensei in a special way. "He wanted to teach us all he knew on the nuclear fuel cycle, more on the philosophy of moving from research to a commercial project." Han was referring to Nakamura's own experience of personally developing fuel cycle technologies at the reprocessing pilot-scale plant at Tokai, but the commercial scale project was transferred to a utility-owned and managed reprocessing plant at Rokkasho with little input from the PNC. The situation was somewhat analogous to Korea when KAERI was very much involved in the commercial NPP system design projects until it was transferred to the industry twelve years later. A similarity between Japan and Korea's nuclear technology development was that both national nuclear labs (PNC in Japan and KAERI in Korea) were responsible for the technical self-reliance drive in the crucial areas (reprocessing for Japan and NSSS system design for Korea). The difference was that PNC was not given a chance at full-scale production before the commercial project was initiated. Korea was fortunate to have KAERI responsible for commercial production for a finite period of twelve years before it was fully transferred to the industry. Being a shrewd manager with genuine foresight, Han must have learned a valuable lesson from Nakamura in holding on to the key technology development in full commercial scale for some period of time. Han said one other important lesson learned from Nakamura Sensei was his strict adherence to and obsession about the "peaceful use only" principle to erase any slight possibilities of proliferation concerns. This was a much more prevailing concern for Nakamura in the plutonium recycle area, which is one of the most proliferation-sensitive technologies.

"Enertopia" man

If Han Pil-soon were the main driving force on the technology side, one person stands out to complement the overall NPP technical self-reliance drive in the management and financing side in the 1980s. He was Park Jung-ki ("Rocky Park"), president of KEPCO from 1983-1987, the most crucial period in the nuclear self-reliance forming stage. Park was born in 1935, graduated from the fourteenth class of KMA (a three-year junior to President Chun Doo-hwan), had a military career as field commander, then served in private sectors as CEO of KHIC (now Doosan Heavy Industry) briefly, then CEO of KEPCO. After retirement from KEPCO, he devoted his energy to promoting sports, particularly unpopular ones like athletic games. He is an active member of the Korean Athletics Association and International Association of Athletics Federation (IAAF). He was the person most responsible for recruiting the thirteenth IAAF World Championship game to Daegu, Korea, in 2011. This biennial athletic event is recognized as the single most important international competition for athletics only. Another aspect of Park's life is as a well-known free-lance writer. He wrote several books in Korean as well as in English. Among the best known are *Plain Stories by a Grandpa (1990), The Art of Leadership (1997), Leadership (2002), The Civil War (2002), Queen of Sports – Tales from Athletic Heroes (2010).* His number-one bestseller, *Plain Stories by a Grandpa*, won a distinction translated into English and was used as a reference text book in leadership and Oriental culture at the US West Point Military Academy.

Park's nuclear connection is believed to be as early as his KMA cadet days, when strategic value and dual faces of nuclear energy and weapons were taught. During his time at KHIC in 1982, he learned the importance of technical self-reliance in design engineering or software power. Being the largest heavy machinery company, KHIC was mainly a hardware factory depending heavily on foreign license design and engineering.

When Park moved to KEPCO in 1983, his "Enertopia" concept was first introduced. It simply meant a country with no difficulties or inconveniences in producing and using energy. KEPCO, being the sole electric utility, is one of the largest corporations and the only owner and operator of NPPs in Korea; Park's Enertopia idea had a direct implication on nuclear technical self-reliance. This nuclear posture was deemed rather unusual for a KEPCO CEO since, as a large nuclear electric utility company, their first priority is always conservatism over innovation to assure power production at minimum risk. This often-times meant KEPCO preferred foreign suppliers for added warranty over local companies, as they had very little records to show, especially in nuclear sectors. This has been a long running dichotomy between KEPCO and KAERI in nuclear power production since one side generally prefers foreign dependence while the other had to fight for equal opportunities for local products and services.

Park was an extraordinary CEO of KEPCO in this regard. His mind was set on nuclear technology self-reliance, and the prime opportunity was coming with the onset of the YGN construction project. In July 1983, Park concluded a KEPCO internal study on a long-term NPP program based on indigenous technology and standardized NPP for cost reduction in copy plant construction. The timing was ripe for KEPCO, with new president Park's grand Enertopia plan and KAERI's Han, with his first success credit for NPP with the Wolsong fuel prototype production. The encounter of two like-minded leaders was bound to create a huge synergy. Park Jung-ki and Han Pil-soon's first meeting came in June 1983, a few days after Han was appointed to KNFC president, only three months after President Chun's visit to KAERI and Park's appointment as president of KEPCO. Since KNFC was a subsidiary company of KEPCO, Park invited Han to his office in Seoul. They had never met before, but both like-mined CEOs felt comfortable with each other immediately regarding their technical self-reliance philosophy, moving on the need to bring well-educated technical staff at KAERI into real KEPCO projects. Park was already informed of KAERI's Wolsong Fuel Project success. Both shared the same philosophy of technical independence, meaning "without technology, you become a slave; only with technology can you act like a master." A new nuclear partnership was building in a way no one had seen in Korea before. Park later served as the KAERI board chairman and member of the Atomic Energy Commission. After all, Park and Han created an era of technology and management alliance in the 1980s when KAERI participated in the NPP construction projects as the fuel and reactor system designer.

Han showed his highest allegiance to Park as his boss, although, technically speaking, each reported to different ministers (Han to MOST, Park to MER). This relationship was deemed an unusual one for typical Koreans, as each ministry normally fought for its own territorial turf. Park recognized the value of supporting KAERI as the center of nuclear brain power and he relied heavily on KAERI for achieving technical self-reliance. On many occasions he made generous financial contributions to KAERI toward building-up infrastructure, such as installing a supercomputer in 1986. His goal of achieving technical independence came with the YGN project together with KAERI and other members of the Electric Power Group Cooperation Council (EPGCC), composed of MER/MOST, KEPCO, KAERI, KOPEC, KHIC, and KNFC. (This topic will be dealt in much detail in Part II of this book.) Park showed his personal confidence on KAERI's manpower as the crucial element for achieving self-reliance sooner than expected with the YGN construction project.[9]

CHAPTER 4:
THE SYSTEM DESIGNER

It took a considerable length of time to learn and appreciate the true value of the NSSS system design for a nuclear power plant in Korea. The gradual learning process of NPP technologies started from operations, civil/structures, non-safety components, moving on to safety grade components, turbine/generators, nuclear fuel, reactor components, then finally the NSSS system design know-how's. The scope of the reactor system design normally covers the reactor coolant loop, which generates the heat from the fuel; auxiliary systems immediately support the reactor coolant loop's normal operation, like the chemical volume control, the safety systems to mitigate any accident conditions, and the main control room design. As such, the system designer defines the reactor type (PWR, BWR, etc.), power level, uranium fuel requirements, radiation exposures, level of safety and accident scenarios, specifications for the major equipment, like the reactor vessel, and the start-up and commissioning test guidelines. Once you master the system design technology, you are able to change the very basic design of a nuclear power plant - for example, to increase the power level and the basic configurations. This is why the system designer assumes the performance warranty of the NPP, prepares the core parts of the Safety Analysis Reports, and carries the main responsibility in the regulatory licensing. Also, the system design technology comes under the export control regime when the technology is to be transferred to a foreign country. It is sometimes considered a "sensitive technology" in a proliferation sense if one can alter the basic functions of a reactor in many ways.

Only four countries developed commercially successful NSSS system designs with "original technologies": PWR by Westinghouse, Combustion Engineering and Babcock & Wilcox, BWR by GE from the US, AGR by the UK, CANDU by AECL of Canada, and VVER by Russia, independently from each other. All three except Canada are nuclear weapon states (NWS). The remaining two NWSs, France and China, had to import the commercial NPP technologies (of which the system design technology is the core): Framatome imported Westinghouse PWR technology in the 1950s; Chinese imported Westinghouse AP1000 and Areva EPR technologies in the 2000s. Japan also had to import PWR and BWR technologies from the US in the 1960s despite their advanced status of basic nuclear technologies. The Korean experience of acquiring the system design technology came in time with the standardization program of the 1980s, but it took longer to prepare the right soil for the new technology to take a firm root based on the commercial NPPs before the YGN episode. Compared to the other nations with the system design know-how's, Korea's story of tackling the challenge came with selecting the right entity to take up the responsibility first.

Mini-standardization

KEPCO initiated the NPP standardization project in the early 1980's, when the first phase of NPPs, Kori 1&2 and Wolsung-1, were being introduced to Korea. Needless to say the phase I projects were all turnkey plants; Kori 1&2 from Westinghouse, and Wolsong 1 from AECL Canada (to some extent Ulchin 1&2 from Framatome). Growing pains of having to cope with three different types of NPPs from three different supplier countries were felt everywhere in the Korean nuclear sector, KECO in particular. Both Westinghouse PWR and Framatome PWR are of the same origin NSSS system; however, many differences still existed in operation and maintenance modes. The Canadian CANDU system design was an altogether different heavy water reactor with lower pressure/temperature requirements. Regulatory licensing requirements were also varied, creating sometimes confusion and duplication between the regulatory body and the operator. If Korea were to have many more NPPs in the longer run (as turned out to be the case), it made good sense to localize the key NPP technologies with specialized Korean entities. In order to achieve this ambitious goal, Korea had to finalize its reactor strategy and the standardization of power reactors to be built in the next generation. Based on the currently commercialized power reactor technologies as of 1980, the NPP standardization study was launched by the MER. Participation entities were KECO, KOPEC, KHIC, KAERI, and KNFC, with KECO as the lead organization. The aim was to reach a long-term NPP construction program with technical self-reliance by the early 1990's and with Korean contents making up 90 percent. Also, reduction in the construction period, as well as initial cost, was envisaged as an outcome of the standardized NPP. In the early 1980's, the level of knowledge in NPP design was rather limited among the standardization participants; KECO with some operation and construction management experience, KOPEC with some architect-engineering plant design experience, and KAERI had only nuclear fuel design experiences. The French model of deploying CP-1, CP-2 series of NPPs by EdF, the electric utility, and Framatome, the nuclear vendor, was bench marked as a national model. The only difference was the inclusion of KAERI, the nuclear national lab, in the study as the French CEA was not directly involved in the commercial NPP planning and design.

The first three NPPs built during the 1970s were contracted on fully turn-key bases from overseas suppliers due to the lack of a domestic technology base. Kori Unit 1, the nation's first NPP supplied by the US Westinghouse turnkey project, went into commercial operation in 1978, followed by the Canadian CANDU reactor at Wolsong Unit 1, and Kori Unit 2 in 1982 and 1983, respectively. Korean contents were limited to civil construction at site and reactor operations by the sole nuclear electric utility company, KECO. The next four NPPs built during the 1980s, Kori Units 3&4 and Yonggwang Units 1&2, were "mini-standardized NPPs" in Korea with the same design and vendors; however, this time KECO took greater responsibility in a component approach to the contracts. All four nuclear islands were from Westinghouse (3-loop

900 MWe PWRs), the turbine generators were supplied from GE, and the architect-engineering was from Bechtel. KECO gained invaluable experience in the total project management while overseeing three foreign prime contractors (Westinghouse, GE, and Bechtel), reaching the goals of schedule, cost, and most of all, quality.[1]

The benefit of equipment localization was greatly enhanced when the foreign supplier was obligated to also provide technology transfer. Many Korean companies first experienced nuclear quality systems as subcontractors to the primes, and gained practical experience and qualifications, such as the ASME N-stamp. More than 35 percent of equipment localization and nearly 95 percent of construction work were carried out by Korean companies in the Yonggwang 1&2 project. The KECO project manager for the Kori 3&4 project, Shim Chang-saeng, recalls: "The most difficult challenge we faced was to bring out the reimbursable type from a fix price contract in the architect-engineering scope with Bechtel without losing the cost control and technology transfer. Without this experience, we could not be successful in the technical self-reliance process with the YGN project." In other words, the experiences accumulated in the Korean nuclear sector, particularly in KECO for its total project management capacity, had reached a sufficient level to set the stage for the next full scope Koreanization of nuclear power technology in the 1980s. Korea Electric Company (KECO), the sole electric utility in Korea, was restructured to Korea Electric Power Corporation (KEPCO) in 1982 by a government decree mainly to address the financial governance issue with minimal impact to the nuclear power projects.

Korea had built and operated nine NPPs before the YGN project started in the mid-1980s; first three units employed a full turn-key approach, followed by the next six units, which employed component approach contracts. All units' design could be categorized as the Generation I NPPs of PWR and CANDU-type reactors originating from the US, Canada, and France. During that period, NSSS vendor companies (like Westinghouse, CE, GE, AECL, and Framatome) incorporated their reactor designs with different architect-engineering companies as

The system designer 55

selected by their electric utility customers. As a result, every NPP design was different. It took the TMI accident and the subsequent slowdown period for new construction to start formulating the user's requirements into a standard design for the entire NPP. This was led by the Electric Power Research Institute (EPRI) in the US. The logical output was the Advanced Light Water Reactor (ALWR) concepts with standardized licensing and an entire plant design approach, or the "customer pull" design rather than the "vendor push" design. This was the onset of the Generation II NPPs, fully standardized for the entire plant design. The YGN project was at the threshold point in the global nuclear power history to take the best advantage of the Generation II approach. Construction of a large number of duplicate plants coincided with the full-scope technology transfer, once and for all.[2]

Reactor vessel only

It happened on a Friday afternoon, only a few months after Han Pil-soon was inaugurated to be the president of KAERI in late August 1984, succeeding President Cha Jong-hee. Han was promoted to the top CEO post only two years after he joined as VP of Daedeok Engineering Center (DEC). He was by then unquestionably a new dark horse on the Korean nuclear scene with successes already demonstrated with Wolsong and PWR fuel projects. Furthermore, Han was appointed to the CEO post of KNFC in addition to KAERI in 1983, indicating strong backing from the Blue House. Han summoned an emergency senior staff meeting (or brain storming) at KAERI's VIP lounge in Seoul. About twenty division director levels and higher showed up from both Taenung campus and DEC, Daejeon. He made the opening statement on the purpose of the meeting: find out the ways and means how KAERI would participate in the NSSS system designer's role for the coming KSNP projects, starting from YGN construction. He also emphasized the growing momentum and confidence gained through two fuel localization projects, and next logical step was to take charge of the entire NSSS system design. Whatever the outcome of this brainstorming session, it would be reported at the next EPGCC meeting at the Ministry of Energy & Resources (MER) to designate local entities for various self-reliance responsibilities.

Heated discussions and brain storming lasted beyond midnight and still no consensus was reached. It was clear not too many senior staff members at the meeting had any idea of what it entailed to become a reactor system designer of a full-scale commercial NPP, as no one in Korea at that time had working experience in the reactor system design. Many doubted whether it was desirable for KAERI to get into the NPP reactor system design business or could be capable of implementing the commercial project responsibilities. The only point they all agreed on was that KAERI had no excess manpower to spare in the new major undertaking, considering the current work load commitments to two ongoing fuel projects. Already KAERI had nearly one hundred staff members committed to the PWR fuel project, in addition to one hundred fifty full time staff assigned to the Wolsong PHWR fuel localization project. As a compromised interim solution, a conclusion was drawn at the meeting at around 2 am for KAERI to take part only in the reactor vessel (including the fuel) design while leaving out the rest of the NSSS design due to a lack of available manpower. Han was clearly not satisfied; however, he did not push at that moment. It was ironic that no one in the meeting that night knew it was not even possible to divide reactor system design into two parts: reactor vessel part only and the rest of the NSSS reactor system. The reactor vessel and the rest of the NSSS are so integrally connected that always a single reactor vendor company (like Westinghouse, Framatome, Combustion Engineering) supplies a system design covering the entire NSSS, including the fuel and the reactor vessel. Alas, it was a simple but too premature reflection of the status of knowledge of NPP reactor systems by KAERI staff at that time in history. It was a mistake to be corrected in the very near future.

The final verdict

Rumor got around in no time that KAERI would not take part in the system design (except the reactor vessel part). Two nuclear entities,

KOPEC, the plant architect-engineering firm, and KHIC, the NSSS component manufacturer, were eager to get into the business as they were better informed about the significance of controlling the system design technology. In fact, they were in competing position with KAERI as to who would get the most coveted nuclear technology sector. A senior confidant of KOPEC's president came down to Daedeok and had private sessions with Han. They were all ready to jump into the system design business and were asking KAERI to loan them qualified manpower, about sixty for KOPEC. Han could not give him a clear answer but was very much puzzled; "How come they want the business so badly without having any manpower while we were declining the opportunity with only a shortage of manpower?" Suddenly a common-knowledge business practice dawned onto Han in that any shrewd businessman would gamble on taking up a lucrative new deal without qualified manpower and relevant technology, rationale being that once you get the business (and the funding), you would recruit all needed manpower and buy technology from foreign sources. That was the simple but hard fact Han and pure scientists at KAERI had no idea about, but that was how most Korean businesses made a record growth in recent decades with all economic ups and downs. That was the difference in mentality between scientists and businessmen. Han made the decision all by himself that KAERI should be the NPP system designer (and not just the reactor vessel designer) for the sake of national nuclear self-reliance, different from other commercial business sectors.[3]

This particular point turned out to be the most difficult issue among the EPGCC members, especially between KOPEC and KAERI. Earlier EPGCC meetings on the subject had KOPEC designated to take up the system design role together with the rest of the NPP architect design, and KAERI to take up only the reactor vessel design (including the core design). KAERI was too timid in showing its true potential, as witnessed by its own brain storming outcome of 1984. A *de facto* split in the NSSS system design between the reactor vessel and the rest of the NSSS by two different entities seemed finished once and for all. Further deliberation, however, began to show basic logical flaws as no nuclear vendor company in the world

(Westinghouse, GE, Framatome, AECL) has such division of responsibilities. In fact, it was rather impossible to split the system design scope into two as such to carry out a real NPP construction project. The EPGCC member companies began to have second thoughts about its practicalities in technical management aspects after seeking advice from foreign consultants. KAERI was the only nuclear entity in Korea with more than three hundred technical staff members in nuclear fuel and power reactor engineering in 1985 with advanced engineering degrees (PhDs, MSs), while other entities had only a handful of engineers of such caliber. One might argue if a NPP construction were the only goal, it would have been acceptable to split the scope, but the YGN project was different. Technical self-reliance was given higher importance over construction, and who should be in charge of the top-tier technology was of paramount importance.

The epoch-making turning point came on June 25, 1985, at a special EPGCC meeting called by President Park Jung-ki at his office in Seoul. It happened to be the thirty-fifth anniversary of the Korean War. All CEOs of the electric power group companies were attending: President of KHIC Sung Nak-jung, President of KAERI/KNFC Han Pil-soon, President of KOPEC Chung Kun-mo, plus President of KPS Kim Sun-chang. The main agenda was to finalize the self-reliance entities for each area of NPP technology in preparation for the upcoming YGN construction project, in particular the ongoing debate on who should take the NSSS system designer's role. Growing confidence and partnership between Park and Han must have played a subtle role in reversing the earlier position taken by EPGCC members. Park commented in his recollection of this point, "It was clear in my mind that the entity with most brain power should take the system designer's role as it is the top technology and able to achieve the self-reliance goal in the shortest possible time."[4] Han also made a strong argument for KAERI to take up the entire system designer's role. The minutes of the June 25 meeting shows this historic decision in much detail. In the technology transfer division of responsibilities for the NSSS, the reactor vessel was now integrated with the reactor system design package assigned to KAERI, while component design and equipment manufacturing was assigned to KHIC. No longer would the reactor vessel and the rest of the NSSS be divided. In the NPP construction contract terms, KHIC was the prime contractor for NSSS supply while KAERI provided full NSSS system design to KHIC. It was truly a day of reckoning for the Koreanization of nuclear power technology, the day of "crossing the nuclear Rubicon" in Korea.[5]

From this day on, KAERI became responsible for technology transfer as well as the supplier of NSSS system design for the YGN project. This historic decision was formally blessed by the 214th Atomic Energy Commission meeting on July 29, 1985. Twelve years later, when YGN construction, together with the technical self-reliance goal, was duly achieved, the entire system design project team at KAERI (grown to three hundred and fifty full time staff members by 1997) was transferred to KOPEC, together with the ongoing NPP system design projects. History will judge that the record number of NPPs constructed in Korea after the Chernobyl accident, and subsequent winning of the UAE contract in 2009, had its crucial turning point on the day of the Korean War anniversary in 1985. The net effect of the nuclear national laboratory holding responsibility for the technology transfer at the same time delivering the system design to NSSS supplier KHIC was to assure self-reliance in the shortest possible time. It was a unique model for a developing country to acquire the top-level technology, transforming Korea into a nuclear export country. From 1985, Korean nuclear entities were lined up once and for all with the following division of responsibilities for respective technology transfer, plus the corresponding NPP scope of supply:

> KEPCO: Overall project management for construction and technology transfer
> KOPEC: Plant architect-engineering
> KAERI: NSSS system design and initial core design
> KHIC: NSSS, turbine/generator component design and manufacturing
> KNFC: Nuclear fuel fabrication

They were the core of the Electric Power Group Cooperation Council (EPGCC) member companies. They were the main pillars of the Korean nuclear power technology for the next decades to come. Every quarter for the next twelve years (1985 - 1997) the EPGCC member companies had the running workshops hosted by each entity in rotation basis to discuss, argue, compromise, and, most of all, learn from each other valuable experiences in different aspects of NPP construction technologies. Building up mutual respect and compatibility in the unique experiences of their lifetime was about to commence.

CHAPTER 5: BEST MENTOR

Getting started

Once KAERI management was assured of the brand-new work assignment of NSSS system design by the highest authority in Korea in July 1985, the first order of business was to set up a new project team called "Power Reactor Systems Division" (PRSD), reporting to VP of Nuclear Projects Rim Chang-saeng. Already the PWR Fuel Design project team was in full operation, implementing joint design activities with KWU at Erlangen, Germany (project manager Kim Si-hwan). BK Kim was drafted from Wolsong Fuel project to head up the new PRSD as the project manager. BK had a mechanical engineering background from Caltech, and worked for NASA Jet Propulsion Laboratory before his return back to Korea in 1975. The project team was assembled and section heads appointed from the best available manpower pool within KAERI with nuclear, mechanical, electronics, and safety analysis backgrounds, as below:

Project management	:	Kim Byung-koo (BK Kim)
Nuclear design	:	Kuh Jung-eui
Fluid systems design	:	Lee Byung-ryung (BR Lee)
Mechanical design	:	Sohn Gap-heon
Instrument and control	:	Hahm Chang-shik
Safety analysis	:	Kim Dong-soo
Project control office	:	Yang Seung-yeong/Lee Ik-hwan

They were all in their late thirties and early forties, with good technical education backgrounds (MSs and PhDs from home and abroad), but had little or no practical working experience in NSSS system design. Nevertheless, everyone knew he was in for the thrilling experience of his lifetime, eager to outperform in full motivation.

The first task for the PRSD team in 1985 was to take part in the NPP standardization project organized by MER together with KEPCO, KOPEC, and KHIC in preparation for the upcoming Yonggwang 3&4 project. Soon enough, pressure was building up to prepare the invitation-to-bid (ITB) documents in achieving the technical self-reliance at the same time as construction of YGN. A total of twenty-three of the world's best known nuclear supplier companies in seven countries were qualified to be on the bidders list:

> NSSS: Westinghouse, Combustion Engineering, Framatome, KWU, AECL, MHI (6)
> Turbine/gen: Westinghouse, GE, BBC, Alstom, GEC, KWU, MHI, Hitachi (8)
> Architect-engineering: Bechtel, Sargent & Lundy, Ebasco, Stone & Webster, Gilbert, Gibbs & Hill, KWU, EdF, AECL (9)
> Fuel fabrication: same as NSSS

KEPCO was in the leading position to select the qualified bidder list to ensure impartiality and no chance of omitting any qualified potential bidders and absolutely assured that all bidders would be given equal and fair opportunities. In accordance with the division of responsibilities (DOR) among Korean entities, each entity started preparing their respective ITB documents: KHIC on NSSS & T/G component design and manufacturing, KAERI on NSSS system design and initial core design, KOPEC on architect-engineering, and KNFC on nuclear fuel fabrication. The total amount of the project cost in building YGN was estimated to be well over three billion USD, including the technology transfer, the largest ever in Korean history. Among the ITB packages, NSSS system design was considered the most important and set the tone of the entire project definitions as the reactor system designer sets the basic NPP type and top-tier character of the plant performance, quality, and most of all, regulatory licensing requirements. The project teams in EPGCC met frequently in the Seoul KEPCO office to harmonize the ITB documents. After three months of working day and night, finally the ITB documents were sent out on November 4, 1985 to all twenty-three bidder companies with a bid document submittal date of March 31, 1986, giving each vendor five months to prepare the bidding documents. KAERI's PRSD project team wasted no time in submitting the ITB document to KEPCO to send out on time. The official language was decided to be English only. A new temporary project office was set up in a remote building, the VIP lounge on Daedeok campus with sleeping quarters, for twenty-four-hour operations mode with special security and confidentiality requirements. It resembled a war situation-room at a military headquarters. It was so different from the rest of the KAERI campus, as if coming from another planet. By then every member of the PRSD project team realized what they were in for, and gradually the importance of the system design work scope, which came to KAERI after such a grueling internal struggle. The NSSS system design ITB specifically called for three separate documents for technical evaluation, plus the price bids in separate volume (the price bid docket was to be opened with bidder's attendance on the final day of the evaluation at KEPCO). The technical bid documents should contain: 1) NSSS joint design process, 2) technology transfer contents, and 3) vendor's NSSS system self-evaluation.

The world's nuclear supplier industry was already in a slow-down mode in the mid-1980s until the fatal Chernobyl accident erupted in the Ukraine in April 1986. It was just one month after the bid document submittal deadline when the project team heard the devastating news for the first time. Then everything changed for the worse. The entire global nuclear sector was entering a nuclear dark age, stopping and cancelling on-going NPP orders. Somehow the bad news did not create a fatal atmosphere in the Korean nuclear community; the on-going bidding process was kept intact. However, many countries proclaimed *de facto* moratorium on NPP operation and construction. Ironically enough, this was truly a blessing in disguise for Korea as it created an enormous "buyers' market" that the nuclear industry had never seen before. Every candidate bidder on the list was more than eager to please Korean nuclear entities in seeking inside information to better prepare their bidding documents. The events that took place during these defining two years (1985-1987), until the contract signing and project commencement, were indeed history in the making. Only twenty-five years later in Abu Dhabi, the Korean consortium was sitting on the other side of the table for signing the UAE NPP construction project as the successful bidder after competitive bidding among Areva of France and Hitachi-GE of Japan and the US.

Bid evaluation

The bidding documents started to arrive at the project office by the end of March 1986, on time as no serious bidder could afford to miss the golden Korean market. They came literally in truck loads in hundreds of special binder documents with multiple copies. Among the six NSSS bidders to whom the ITB was sent, KWU of Germany and MHI of Japan declined to submit the bid for reasons of their prerogative (some believed Mitsubishi had difficult licensing conflicts, with Westinghouse being their original technology holder and competitor). The remaining four (Westinghouse, Combustion Engineering, Framatome, and AECL) were in for a fierce competitive battle to win not only the YGN project but all of the future Korean NPP market. It was a matter of life or death as to who would win the contract. The future of NPP technology transfer to Korea was to be decided among the four global NSSS vendor companies.

The bid evaluation criteria was carefully screened and decided at KAERI for NSSS system design, and kept in a sealed envelope in a secure storage safe just before the ITB documents were sent out in November 1985. This was to ensure absolute fair competition when the bid documents were in. The bid evaluation criteria were detailed enough to generate qualitative numbers in comparing four bidders. Each bidder's name was coded into "가", "나", "다", and "파" (Korean alphabets) to reduce any biases from each evaluator. It took three months' time (day-and-night) working together in the war room for twenty KAERI evaluators to produce the final technical evaluation summary in three volumes (Joint Design process, technology transfer contents, and vendor's NSSS system technical description), comparing four vendors (Westinghouse, CE, Framatome, AECL).[1,2,3] By July 10, 1986, KAERI's final evaluation report was ready for submittal to KEPCO. It contained a surprise choice of the "successful bidder" recommendation as the unanimous conclusion from the twenty KAERI evaluators' final scores. It was clearly "CE" over Westinghouse, Framatome, and AECL, judged by the purely technical scientists at KAERI. Like-minded people must have thought similarly, as the eagerness and willingness to transfer the utmost technology know-how's to Korea was most evident with CE over the other three bidders, thus obtaining the highest evaluation scores. The outcome was the most sensitive top-secret information no one was allowed to even talk about. Once the final report was submitted to KEPCO on July 10, 1986, the entire evaluation team took a special summer holiday leave to recuperate from the intensive stress accumulated over a year and to get away from the tension and secrecy. No one believed, however, that their favorite son would win the contract.

"Surprise" announcement

It took another two grueling months before KEPCO made the final announcement of the successful bidders. They had to integrate other areas' bid evaluations (architect-engineering by KOPEC, NSSS/TG supply by KHIC, nuclear fuel fabrication by KNFC), plus the crucial price bid information. Proportional to the overall project cost (around four billion in 1986 dollars), which was earmarked to be the single largest construction project in Korean history, politically savvy KEPCO top management had to have prior clearances from the top-level government officials, it was presumed.

It was in the morning daily newspapers of September 30, 1986, in Seoul that the world first discovered the final outcome of the YGN NPP international tender, as the normal practice of KEPCO for such occasions. Successful bidders first to negotiate the supply contract were: **for NSSS system design and technology, Combustion Engineering (CE) of USA; architect-engineering, Sargent & Lundy (S&L) of USA, and turbine/generator technology, GE of USA.** NSSS supply, initial core design, and fuel fabrication technology were also awarded to CE as a necessity to link with the NSSS system design. The nuclear vendor companies (Westinghouse, AECL, and Framatome), who had well-established Korean market share, all lost out in this bid of a lifetime. Particularly for Westinghouse, the shock of losing the future Korean market altogether must have been a startling news. They had already supplied six PWR units (Kori Units 1-4, Yonggwang Units 1-2) as turnkey/component projects and there was much speculation that Westinghouse was the sure winner, considering all their built-in advantages and internal connections in Korea. Also, S&L for AE scope was a surprise to many. Bechtel was expected to win the AE contract since it was the largest AE company in the projects for the existing NPPs. Considering the fact that Combustion Engineering had no prior NPP contracts in Korea, nor any technology transfer licensing experience abroad on the NSSS system design technology, the outcome of the tender did make headline news in the global nuclear scene in 1986. It was the contents and enthusiasm of the technology transfer by CE that impressed the Koreans most over Westinghouse. One interesting anecdote on this point was delivered at an ANS-AIF luncheon speech in Washington, DC in November 1986, just after the Yonggwang announcement by Lee Chang-kun (who later served as a Korean Atomic Energy Commissioner and a well-known writer in Korea). Lee was paraphrasing from Shakespeare's *Julius Caesar* to lament the downfall of the Westinghouse hegemony in the Korean nuclear scene:

> *Who is here so vile that will get rid of*
> *The mutual cooperation and technology transfer?*
> *To him the poor and weak Koreans will sing in chorus*

> *"Auld Lang Syne"**
> *I will not pause for reply.*
> (* Scottish poem by Robert Burns, set to the tune of farewell folk song)

Then the president of CE Nuclear, Shelby Brewer, was quoted as saying that he liked Lee's poetry so much as to have it framed on his office wall. No doubt under Brewer's guidance, the CE team was able to offer an extraordinary technology transfer package to win them the contract, and ultimately helped Koreans to reach their goal of technical self-reliance.[4]

One sour loser nuclear vendor company (you can guess who) even filed a protest in the media demanding clarifications on how KEPCO made the bid evaluations, claiming the winning CE's 1,300 MWe System 80 design had to be scaled down to 1,000 MWe, thus had unproven safety. The smallest and little-known nuclear NSSS vendor CE company was about to put their name on the world nuclear map. Only twenty-three years later in Abu Dhabi, history repeated itself with a similar nuclear surprise announcement from the UAE's largest emirate, awarding KEPCO as the successful bidder for supplying four units of 1,400 MWe PWRs from Korea in competition with Areva of France and Hitachi/GE of Japan/USA. The selection of the virtually unknown global nuclear vendor KEPCO in 2009 had much resemblance with the awarding CE in 1986.

Global nuclear industries went through dynamic merger and acquisition exercises in the past three decades to survive the low-business post-Chernobyl era. CE was later merged with its archrival, Westinghouse, then subsequently merged with Toshiba of Japan.

Technology over politics

Going back to the historic decision to take CE for all future commercial NPP technology in Korea, historians may argue the validity of such a case in the coming years. What became clear as dust settled down was, KAERI's recommendation after the technical bid evaluations was accepted in totality by KEPCO and the Korean government without alteration. Indeed it was a surprise even for local Koreans (including myself) at that time that KAERI's mere technical evaluation results were honored. With the political climate of Korea in the mid-1980s, as in most developing countries, it was natural to expect a political deal in a mammoth national project like the YGN construction contract. This automatically means the final successful bidder is normally picked by higher-up political considerations rather than by pure technical evaluation results like KAERI's, merely being used as a cover-up to legitimize the evaluation process. The question of how the final selection was made, indeed free from political interaction then in 1986, was of much mystery a quarter century later, now that the technology selected then resulted in a huge success in domestic NPP bases, as well as winning the first overseas NPP export contract in 2009. Detailed truth may never be fully disclosed for lack of hard, documented evidence on such a typically political scenario. However, I am willing to make a case of an honest-to-goodness conclusion for the YGN nuclear project being free from political decision behind the scene based on the following facts:

1. The original budget of $3.4 billion for the YGN project to build and complete technology transfers was achieved in-budget and on-time. Most national mammoth projects experience cost overruns together with construction delays.
2. Subsequent KSNP repeat projects built in Korea achieved lower capital costs and shorter construction periods, taking advantage of learning experiences. This is the basis for the Korean NPP's international competitiveness.
3. Personal interview with Park Jung-ki, president of KEPCO in 1986, confirmed his conviction to support the KAERI recommendation for selecting the best reactor vendor for technical self-reliance without any political considerations.[5]
4. A massive government prosecutor's investigation of the YGN project implementation in 1989 concluded a 'clean bill of health' after summoning some 156 witnesses from EPGCC

company's key project staff over six months. They were focusing on the possibility of political slush funds, as alleged by some opposition party politicians.

It was a rare case in our modern times that the final decision on a multi-billion-dollar contract involving foreign and domestic companies was based on technical/economic merits alone rather than hidden political agendas in a developing country like Korea in the 1980s. As a Technical Cooperation (TC) director at the IAEA, I have come to learn the reasons why so many developing countries that started ambitious nuclear power program ended up with failures. To formulate meaningful TC projects in Member States, we must prepare the Country Program Framework (CPF) to analyze the basic nuclear history of all projects and reasons behind them. I was surprised to find out, almost without exception, the NPP construction deals were made between the host politicians and the nuclear vendors to influence the bidder selection and project implementation. Rampant cost overruns and project schedule delays were common, without bringing energy independence nor nuclear technologies to the host country. It was sad stories overall, simply reflecting political realities in those developing countries. However, the South Korean case was different. Technology was ahead of politics for the YGN project. How could it be possible? After the investigative work in writing this book, I felt that the 5th Republic under President Chun might not seek political fund-raising from big projects like the YGN. What is important, however, is the fact that the nuclear projects in Korea were free from political fund-peddling, unlike in most other developing countries. This could be the single most important reason the Korean success story became reality. Once again, the South Korean nuclear program was blessed from the nation's highest office in its unique way.[6]

The selection of CE for the NSSS technology partner in 1986 did turn out to be a happy marriage for both Korean and US counterparts. It was a true blessing for Korea. CE had one of the most advanced NSSS designs, the System 80, but more importantly, they had a willingness to transfer the full spectrum of commercial reactor technologies once the recipient part earned the credibility. In fact, it was up to the Korean entity's ability to absorb new technologies in designing and manufacturing the reactor system software and hardware. A good mentor and mentee relationship was about to begin.

Contract negotiations

Once the initial excitement and confusion over the successful bidders announcement in September 1986 was over, the next serious business step of drafting the contract documents took place in Seoul. By then each Korean entity had its US counterparts identified in the successful bidder list for separate contracts preparation. For the first time in Korean nuclear history, KEPCO would sign the NPP construction contracts with five Korean prime contractors, who would sign subcontracts with the successful US bidders (in parentheses below):

Nuclear Fuel:	KNFC/KAERI (CE)
NSSS:	KHIC/KAERI (CE)
Turbine/generator:	KHIC (GE)
Architect-Engineering:	KOPEC (S&L)
Construction:	Hyundai Construction

Overall contract negotiation was about to begin with the new KEPCO counterpart changed from the New Project Team under Lim Han-kwae (a Yonsei University graduate in physics) to the Nuclear Construction Division under Shim Chang-saeng (an SNU graduate in mechanical engineering). Shim was a veteran nuclear project implementer at KEPCO, demanding many new features and clauses to be incorporated into the contract documents. Jim Veirs, CE project director, thought it would take a week or so before the final signature. Once the negotiation began under Shim's leadership, it took more than six months before the final contract documents were signed. It was painful to go through the rigorous clarity of any ambiguous terms and conditions in the contract at that time. However, thanks to this thought-provoking negotiation process, very few contract disputes occurred during the entire project life cycle.

A distinct feature of the YGN contract scheme was to include technology transfer contracts in parallel with the supply contracts. Overall, the NSSS contract scheme showing contracting parties (KEPCO, KHIC, KNFC, KAERI and CE) with two prime contractors to supply NSSS (KHIC) and the initial core fuel (KNFC) to KEPCO, while KAERI was supplying system design to KHIC and initial core design to KNFC. KAERI in turn had software design under the Joint System Design/Initial Core Design contract with CE, while KHIC had a direct component design and manufacturing contract with CE, and KNFC had a separate fuel fabrication contract with CE. In addition, two parallel technology transfer contracts were signed between KAERI/CE and KHIC/CE for system design, and component design and manufacturing technologies, respectively.[7]

Six months of intensive contract negotiations followed before the formal contract signing on April 9, 1987. Altogether twelve supply contracts plus three technology transfer contracts were first signed between KEPCO and five Korean prime contractors (KHIC, KAERI, KOPEC, KNFC, Hyundai), then between the Korean prime contractors and US subcontractors (CE, S&L, GE), all in one seating at the KEPCO head office in Seoul. All contract documents were in English and also separately in Korean versions. A special clause was inserted in the General Terms & Conditions stipulating when the interpretation of contract clauses were different among the parties, then the Korean version would prevail over the English version. It was the first time in its history that KAERI became a contractually responsible, legally binding party in a commercial NPP construction project. Among the highlights in the NSSS contracts were:

- Delivery warranty: penalty payment assessed against late deliveries, CE responsibility
- Performance warranty: CE would bear penalty payment for continuous electricity output. From the next NPP project, KAERI would bear this warranty clause

Joint System Design: KAERI designers moved to CE Windsor office first, then the Joint System Design work would move back to Daedeok. CE would bear the design output responsibilities

Grant of License: royalty-free use rights for future Korean NPPs as well as overseas export projects (after 10 years)

Chernobyl fallout

In the early morning hours of April 26, 1986 at the control room of Chernobyl nuclear power plant unit 4, located near the northern border town of Prypiat in Ukraine, Soviet scientists conducted an unusual experiment of a sort which was later known as the Chernobyl accident, the worst nuclear accident in history. In addition to immediate casualties at the site, westwards wind blew radioactive clouds west of the site, bringing heavy fallout to Belarus, Germany, and Scandinavia. International response was almost horrifying; public perception on anything nuclear turned from worse to worst. Major industrial countries with running nuclear power program were reconsidering their nuclear power program, without exception, from immediate safety evaluations of their operating fleets to moratoriums and outright cancellations of any ongoing construction projects. For the next twenty years or more, most nuclear power countries in America and Europe experienced downsized nuclear power program until recently. Global nuclear industries virtually went into a long dormant mode while young and able manpower stayed away from nuclear sectors. South Korea was perhaps a unique exception as it was in the middle of a national NPP technology localization drive with the bidding and selection of technology partners announced in 1986. Most fortunate for Korea was the fact that the westward wind from Chernobyl did not bring any radioactive fallout on the peninsula. There was no public fear or outcry against nuclear power at that time. The Korean government quietly conducted situation review and concluded to proceed with the YGN project as planned, recognizing any delays in the construction of the next series of NPPs would seriously jeopardize the economic development program. This was a true blessing, in retrospect.

The Korean side benefited immensely from drafting exact contract wording at the time of the contract negotiations. It was only several months after the Chernobyl accident in April 1986. Unusual clauses like "royalty-free condition for export to a third country (after 10 years from the technology transfer)" and "subcontractors bear warranty clauses" reflected the total buyer's market condition during the negotiation period. By the time full sets of YGN contract documents (in English as well as in Korean) were duly signed and went into force, they contained many clauses that were unthinkable before, but now favorable to the Korean side, such as the warranty clauses and royalty-free clause for export case, dispute settlement clause, and many others. Both the US and Korean sides thought the probability of having a Korean NPP exported to a third country was extremely small, but under the Korean side's insistence, CE had to agree. When the news of UAE export contract was out in 2009, it was time to rake the benefit of royalty-free conditions. It is ironic that the Korean nuclear power technology owes a lot to the Chernobyl accident - a classical example of turning crisis into opportunity.

The project implementation started on time as soon as the government approval on financing arrangements was agreed, following the contract signing. As of May 1987, all project signs showed the implementation was off to a good start. But soon enough, that was not to be the case.

CHAPTER 6:
AFTERSHOCKS

The magnitude and impact of the Yonggwang bidders' selection in 1986 was bigger than anyone had anticipated. The total project cost was estimated to be a record high of 3.4 trillion Korean *won* (about $4.4 billion), the single largest project in history. More important was the fact that the losers would be virtually cut off from all future NPP construction business as the YGN was expected to be the lead plant for the fleet of future standardized NPPs in Korea.

Come 1988, the new president of the Republic of Korea, Roh Tae-woo, was inaugurated after a national election to succeed Chun Doo-hwan's administration. The most urgent agenda for the new president was to manage the Seoul Olympics in September 1988. It turned out to be an epoch-making event for Korea to put its name on the global map. What no one was quite ready to accept was the rapid democratization process that followed immediately after the Olympic closing ceremony. Overall, the democratization process was a long-awaited welcoming trend at that time, following nearly a quarter-century of military governments (eighteen years of Park Chung-hee, plus seven years of Chun Doo-hwan regimes) in Korea. Rapid economic developments were achieved with the highest national priority at some cost to democratic processes like human rights, as we know today. The YGN nuclear project, which was in its first year of implementation, fell victim of this democracy movement of 1988. Major public projects decided during the past administration were put under rigorous public scrutiny, as no one had ever seen before. The largest single project in Korean history was a natural target for investigation for any wrong-doings. This brought several aftershocks none of us in the Korean nuclear sector had anticipated, and were thus totally unprepared for. One was the civil protest from a bid losing company, Westinghouse Electric in particular, for nearly two years after the bid was awarded to Combustion Engineering. The other was the prosecutor's inquiry into the ethical validity of the YGN project, in the next two years immediately following the Seoul Olympics, which stemmed from the accusations made by the losing company. News media and the National Assembly audit hearings from both sides of the Pacific had hundreds of articles on accusations, rebuttals, and public hearings, including the prosecutor's investigation results.

Looking back now some twenty-five years later, the situations were representative of the changing times in Korea during the 1980s period, regarding the award of such large public projects, and the intimidating methods fearlessly employed by the well-known foreign mega- company. The way that the matters were handled by the Korean nuclear entities (primarily KEPCO and KAERI) for the YGN project was truly epoch-making in a new approach with technology first. It can be said clearly now that the winners selected were based on the most qualified to meet the Korean project goals, to meet the construction delivery, as well as the most favorable technology transfer to the self-reliance goals. The losers must have underestimated the national determination and level of the technology transfer conditions. With all the external distractions, it is worth noting that the construction and technology transfer projects got started on time, progressed satisfactorily, and were little interrupted by it all. My personal experiences have been that a mega project that makes it through the first two years successfully will likely end up successful in the end, – and vice versa – those that get in trouble early never quite recover the schedule delays and cost overruns. So it is with much relief and a smile to look back on those first "aftershock" years of agony, hard work, and damage control by the project management team from both sides of the Pacific. It paid off handsomely. Not only the construction project and technical self-reliance reached the original goals on time and in budget, but, more importantly, the external probing helped strengthen the project's internal fabrics and increase determination to the end.

Sore loser

Immediately after the successful bidders were announced by KEPCO on September 30, 1986, the first surprise of the YGN project was a series of public media campaigns launched by Westinghouse, who was the principal loser in the bidding, which lasted the next two full years. One of the most respected US electric companies, and a longtime nuclear business dominant in Korea, which had been expected by most to win the bidding, the 'big circle W' company seemed to have made it a corporate policy to get the successful bidder decision reversed. Their main focus was on the "proven design" of the YGN reactor system design, namely the scaled-down hybrid of the System 80 design (1300 MWe, or 3817 MWt size) and the ANO-2 core design (1000 MWe, or 2825 MWt size). They claimed the reactor system design was unproven and thus unsafe, so the Combustion Engineering bid should have been disqualified from the beginning. Literally speaking, the "proven design" must be identical to other NPP existing elsewhere in size and configurations or "replication," at least in the reactor system design area. Although replication may have value, it was never the intent of KEPCO or the ITB, as it retards technological advances and, thus, Korean participation. Critical decisions were already made even before the bid evaluation that the project would take the most up-to-date technology within the "proven design" categories, as it would be the standard design for future NPPs in Korea. It was a delicate choice between the "up-to-date" versus the "outdated" technology, and the "proven" versus "unproven" system design, since every NPP design is different and only a few new nuclear builds were available anywhere in the 1980s.

Only a few weeks after the bid announcement in 1986, Westinghouse submitted the protest docket of over one hundred pages to KEPCO on the "proven design" issue. KEPCO and KAERI reviewed the concerns raised with its selection of the NSSS vendor and concluded that no new issues had been raised that KEPCO/KAERI team did not consider in its original evaluation. As an additional proof of the "proven design," CE also submitted the USNRC letter dated November 20, 1986, from Harold Denton, director of Office of Nuclear Reactor Regulation. In it

Denton stated, "As a result of our review, the staff believes there is reasonable assurance that the proposed 2825 MWt version of the System 80 design, if properly implemented, could satisfy the same regulatory standards applied to the CESSAR Final Design Approval." In other words, USNRC would give reasonable assurance that the scaled-down hybrid version of the System 80 design, as being built in Korea, would be licensable as it were for the full-size System 80 plant in the US. The only caveat was "if properly implemented," meaning constructed with full quality assurance according to the design specifications. While this was convincing enough for the Korean project team, apparently it was not for Westinghouse. On February 3, 1987, president of Westinghouse Robert Kirby sent a protest letter directly addressed to the President of the Republic of Korea, Chun Doo-hwan, at the Blue House (the official presidential residence in Seoul). Kirby wrote,

> "... The risks I am referring to here not only include the risks involved with safe operation, but also include the cost risks associated with redesign, modifications and schedule extension. Should there be any problems with the implementation of an unproven design, it would be very difficult to explain, after the fact, why that design was chosen...."

Bluntly speaking, Kirby was telling the head of a sovereign state that "you Koreans didn't know what you were doing, so you better listen to me now..." When I first heard about this letter then, I thought it must be in the best form of Westinghouse's arrogance which lent insult to the injury. Three weeks later, a response letter from the minister of Energy and Resource Choi Chang-nak was sent to Kirby, stating,

> "With all its ramifications for the future course of our power sector, we have been most careful, through KEPCO, in selecting right system, right not only in economics but also in technology and safety, those aspects you have expressed your concern. And I have every reason to believe that KEPCO

succeeded in following this guideline, including in the selection of successful bidder for the NSSS."

Looking back on this episode a quarter of a century later, one cannot help but feel how fortunate we were in selecting CE over Westinghouse. It was gratifying to see the strong vote of confidence from the top of the Korean government on KEPCO's decision, which was permeated from KAERI's recommendation. I wonder if we were to have selected the Westinghouse technology, then where our level of technical self-reliance could be today. Considering the arrogant nature of the 'big circle W' mentality imbedded in their corporate culture, it would have taken far more effort, if not have been impossible, for Koreans to get to the core of the core technologies. It also taught us an unforgettable lesson: the importance of the NSSS system design technology, which was so central to the controversies involved. The fact that the least commercial but most technical Korean nuclear entity, KAERI, was responsible for the system design technology with its mentor and technology counterpart, CE, turned out to be a true blessing. In the end, Westinghouse tried to politicize the issue and this caused additional burdens in the next two years by the prosecutor's office, as the opposition party was eager to generate a big scandal case that did not succeed.

"Technology over politics" is the main overriding theme in this book to tell the world how the Koreans managed difficult corners: in selecting the PWR reactor type, in designating KAERI to the system design, and in selecting Combustion Engineering. The Westinghouse intervention finally calmed down by the end of 1988, but yet another unexpected turmoil was waiting for the project, this time from the domestic political scene.

Prosecutor's inquiry

The Korean national assembly initiated the audit investigations on major civil rights issues and national projects in October 1988 for the first time. This audit investigation was imbedded in the national constitution but never exercised in the past twenty years for the sake of economic developments. The Economic-Science Committee of the National Assembly selected the YGN project for in-depth investigation. Opposition-party assemblymen, including Hwnag Byung-tae and Cho Hee-chul, were seeking allegations on how the successful bidders were selected to be the suppliers of YGN. One of their main arguments was that CE's NSSS design of 1,000 MWe NPP had never been built anywhere before, thus claiming the lack of proven safety. They seemed to have taken the same "proven design" argument from the Westinghouse allegations. They coined a new media term, "scaled down patch-up reactor," to discredit the YGN design. Between the lines they were alleging possible bribery case of the past administration in favor of winning the contract. At one point they were demanding KAERI to obtain a licensing certificate from the USNRC to assure the nuclear safety of YGN which was still in the design stage at Windsor, Connecticut. All efforts by KAERI and KEPCO project managers to explain and convince the law-makers on the inherent safety margins in the Korean regulatory process turned out to be to no avail. Their minds seemed to be set on staging a major corruption scandal, which would certainly become a political hot potato and embarrassment to the previous administration. The national assembly decided to put this issue to the prosecutor general's office in the Ministry of Justice for a special investigation in November 1988.

The prosecutor's office under chief prosecutor Chung Sang-myung spent nearly six months calling witnesses to verify the allegations. Over one hundred and fifty project-related staff members from the EPGCC companies, from the CEO president to working-level staff, were summoned separately to the prosecutor's office in Seoul. In particular, the project management office staff of KEPCO and KAERI was the prime target, as under KAERI's recommendation the successful bidders were selected. What started out to be a national pride and joy project was about to become a national disgrace. At the end of six months, the special prosecutor Chung made a final conclusion on the allegation: "All project staff investigated showed no sign of inconsistency," thereby concluding the YGN project to be free of any wrong-doings. It was a welcomed relief that all of us could now go back to the normal work of designing, learning and equipment manufacturing. The project's honor was saved at last by mid-1989.

Financial and legal allegations may have been lifted by the prosecutor's final verdict. However, lingering doubts and the validity of the NPP technology self-reliance drive were in many law makers' minds as they had little idea of what the NPP technology transfer was all about. The total price tag of 3.4 trillion *won* was about 20 percent higher than that of the previous project, Yonggwang Units 1&2. This was enough to activate suspicions among some politicians about bribery and embezzlement charges to select an "unproven design" by CE. KAERI sent the first batch of joint design team members, forty-eight engineers, to the CE head office in Windsor, Connecticut, in December 1986, even before the contracts were signed. Soon one hundred and twenty Korean project staff members were stationed in Windsor to carry out the initial phase of the joint design at the CE home office, side-by-side with the CE system design team. Their mission was to complete the major design specifications by December 1989, when the design center was to be relocated to the Daedeok KAERI site, according to the pre-agreed "Transition Plan." When the prosecutor's investigation was underway at home, the joint design process was making good progress under strong teamwork already developed among US-Korean project members at Windsor.

Several leading national assemblymen from the opposition party visited the Windsor site to witness the real NPP design work in progress for investigation purposes. The visiting team, headed by national assemblymen Lyu Jun-sang and Cho Hee-chul, was escorted to the Windsor design office in September 1989. Kang Bak-kwang, then the science attaché at the Korean Embassy in Washington, joined the visitors from New York's JFK Airport. They had informal open discussions with the CE management, then more questions for the KAERI design team, working side-by-side with CE engineers for two days. Kim Jin-soo, the site manager at the KAERI Windsor Office, was effective in presenting the progress of technology transfer and the proven safety of the

YGN system design to the visitors. This visit must have played some role in converting the negative thoughts on nuclear projects in general, and alleged suspicions of the YGN project in particular, for the opposition party members like Lyu. Kang recollects the visit of the hard-core opposition leader to Windsor: "Mr. Lyu almost went into tears after eye witnessing KAERI engineers at work, saying, 'They are the true patriots' of nuclear technology self-reliance." Hundreds of hard-working, well-respected Korean engineers at work together with CE counterparts in mastering the core technology must have changed an old politician's fixed mindset.

Two months after their return back to Korea from the trip, the opposition party leader Kim Dae-jung (who later became the president of Korea in 1998) paid a visit to Mokpo, his hometown, on November 28, 1989, where Kim commanded overwhelming support. He made a public statement when asked by locals on the validity of YGN nuclear projects, saying, "Energy-resource-poor countries like Korea have no other alternatives but to rely on nuclear power, but need full consultation with the neighboring communities." This was the first positive statement from the opposition party leader on nuclear power, known otherwise as the "Mokpo Declaration." Windsor trip reports from one of Kim Dae-jung's confidants may have played a positive role in reversing the opposition party line policy on nuclear power. It did become apparent throughout the last half-century of Korean governments, conservative ones as well as liberal, that nuclear power enjoyed special privileges, in order for Korea to become the world's fifth largest nuclear power producing country.

CHAPTER 7:
LESSONS LEARNED THE HARD WAY

The unexpected episodes, or the "aftershocks," immediately following the successful bidder announcement in 1986 did serve as wake-up calls to the Korean nuclear entities as to the socio-political aspects of the mega nuclear project, in the short-term through the loser's protest and the prosecutor's inquiry. The project team did cope with and survived the situation well enough that the overall construction and technical self-reliance programs were proceeding without a setback. However, much bigger and far-reaching impacts to the nation and the nuclear project in particular were looming from the end of the 1980s, following the Seoul Olympics in 1988. Perhaps because the speed of economic development came in such a compressed time-scale in an overpopulated country with virtually no natural resources, perhaps because we always had a hostile neighbor in the north, the Korean nuclear program underwent a totally unexpected course of events that no one had anticipated. Two historic nuclear events need to be better understood, as the consequences of these are still very much alive and ongoing as I am writing this book. One was the radwaste disposal program amongst the anti-nuclear movements, which became the top national issue over the past two decades, and the other was the KEDO LWR project and nuclear armed North Korea. Keep in mind that the past half-century of economic growth came in parallel with the democratization of South Korea in all social strata, perhaps with too much democracy erupting at times. Nuclear projects in Korea had to go through the midst of the democratization process in the South and nonproliferation process in the North as if it were a given destiny.

NIMBY to Gyeongju

The Korean nuclear community enjoyed relative public support until the 1988 Seoul Olympics. By the mid-1980s, Korea already had operating, or was building, nine units of NPPs at four different sites along the coastline: Kori, Wolsong, Ulchin on the east coast, and Yonggwang on the west coast. In addition, the nation's most ambitious NPP technology self-reliance drive was underway with the onset of the YGN project at Yonggwang. The general public had not much say or chance to speak out on big national project decisions like NPP construction as the country was under military regimes of Park Chung-hee and Chun Doo-hwan for over twenty years. It was a national symbol of economic progress and pride delivering essential electrical energy for the fast-growing heavy industries and booming export-oriented economy. Maybe it is fair to say that South Korea was undergoing its formative stage of socio-economic framework, setting up its basic infrastructure not to be tampered with over civil/social issues. This provided a golden opportunity to set up the main nuclear infrastructures like the nuclear laws, R&D infrastructure, and electric utility to build and operate NPPs at four sites together with ample electrical grid networks without any social resistance. Economic developments had higher priorities than democracy development, to come out of hunger and poverty first. Labor movements with organized unions virtually did not exist until 1988. During the 1970s and 1980s, promotion of nuclear power was planned and implemented by the energy ministry MER, and KEPCO, with technical support from MOST and KAERI. The concept of nuclear safety and independent national regulatory was only beginning to form at KAERI as on-going NPP projects were turn-key ones from the USA, Canada, and France, with their own regulatory frameworks.

The onset of NPP technology self-reliance and the political democratization process in Korea coincidentally happened immediately following the Seoul Olympics in 1988. Collision between the YGN project implementation and the birth of the anti-nuclear movement was almost inevitable. It all started out with a minor incident at the Yonggwang NPP site, only fifty kilometers west of Gwangju, the provincial capital of

South Cholla Province. The general public still remembered vividly the bloody civil uprising occurred in Gwangju on May 18, 1980, against the newly formed military government. A fishermen's compensation protest at Yonggwang in 1987 had its roots in deep mistrust of any government projects, particularly in the South Cholla Province with its long history of discrimination and civil unrest. Local residents near the Yonggwang site started a protest demonstration seeking just compensation of fisheries at the beginning, soon to change its character to become the first anti-nuclear movement in Korea by 1988. They formed a coalition of environmental campaigns with grass-root support from local religious groups, civil rights groups, and newly formed labor unions. One Catholic priest at Yonggwang parish became active in the anti-nuclear movement through his sermons that nuclear power was against God's will in tampering with the atoms. As irrational as it may sound, the anti-nuclear movement was gaining momentum appealing to public fears of nuclear nightmare scenarios like birth defects and cancer increases allegedly caused by radiation releases from NPP operations. The main entrance to the Yonggwang NPP site became a favorite demonstration venue for the anti- nuclear protesters demanding cancellation of the new YGN NPP project. Local police and security guards were sometimes overwhelmed by the protesters. This movement spread like wildfire to other NPP sites: Kori, Wolsong, and Ulchin. The YGN project teams, including myself from KAERI, and KEPCO had to attend numerous public hearings and debates in Seoul and other provinces with NPPs, often times facing emotional and hostile audiences to defend nuclear safety and contributions to local economy.

It was the beginning of real public acceptance of nuclear power in Korea that the project team was least prepared for. In retrospect, the YGN project not only initiated the NPP technology self-reliance but also ushered in the public acceptance of nuclear power in dealing with local populations and the general public for the first time in Korea.

Anti-nuclear movements against electric energy had its limitations for the public, in that overwhelming benefits from nuclear energy, plus economic gains for local communities, were undeniably huge, provided that safety is warranted at any cost. However, when it came to the radwaste disposal site issue, South Korea discovered a totally different and hostile situation, as in many other nuclear power countries. In the late 1980s, the Korean government started a serious campaign to designate candidate sites for disposal of all radioactive wastes coming out of the large number of NPPs in operation, under construction, and in future plans, plus radioisotope wastes from medical and industrial uses. The Radwaste Fund was created in 1989 to accumulate sufficient funds for the final stage of the nuclear fuel cycle to radwaste treatment, disposal, and decommissioning of nuclear facilities under the "polluter pays" principle. KAERI was designated as the national lead center for technology development as well as the radwaste disposal project. It was Han Pil-soon's last ambitious project in establishing the back-end fuel cycle technology together with the radwaste project. Han had every reason to believe his last wish would come true based on the successful track record of projects KAERI managed under his leadership. The National Assembly passed special legislation to levy 1.865 Korean *won* per every nuclear kilowatt-hour generated by KEPCO. Nuclear power appeared to have a sailing start up until the radwaste disposal site issue became open to the public.

Timing of the dramatic events leading to the selection of the site coincided with the rising of anti-nuclear movements in 1989. MOST was designated to manage the Radwaste Fund, thus the mission to find a suitable site for the permanent disposal (initially one site for low and medium-level wastes plus the spent fuel) fell on the shoulders of the scientific entities. This turned out to be a serious mistake. Perhaps naive at best, site after site, candidates failed to win popular public acceptance for the waste site, an issue that lasted nearly the next twenty years, costing several ministers' dismissals. The first candidate site investigated was Yeongdeok in 1988 on the east coast, moved to Anmyon Island in 1990, then to Guleop Island on the west coast in 1994. With one site after another, each time the government had to cancel the whole operation following violent protests from local and anti-nuclear interveners. Dismayed by the poor progress, the government decided to transfer the entire radwaste site program out of KAERI's hands to KEPCO, together with the reactor system design and nuclear fuel projects in 1997. KEPCO's approach was to convince the neighboring NPP site communities (Ulchin, Yeongdeok, Yonggwang, and Gochang) to accept the radwaste repository under voluntary basis as they were more familiar with nuclear issues in general. This initiative also ended miserably when one of the local district chiefs tried to convince the population too aggressively, which caused public riots of protest, to no avail. NIMBY ('Not in my backyard') sentiment was prevailing, gaining momentum from one site to another.

Finally a totally new approach was adopted under President Roh Moohyun's administration by passing a new special law on the "community support legislation to the inviting radwaste disposal site" in 2005. The new law separated out the most controversial issue, that is to limit the site only for the low/medium-level wastes, and the high-level wastes from the spent fuel were to be excluded. A deep geological rock cave repository method was proposed in order to soften the public opposition, despite the high costs involved. This was contrary to other nations' low/medium-level repositories, which are generally of land-fill type. Moreover, the government made offers on huge economic development incentives to the inviting locality, and would seek voluntary, competitive applications from local governments that could demonstrate the will of the majority by referendum results. This new law finally made the turning point. Historic local referendum voting was held at four applying localities on November 2, 2005: Gyeongju, Yeongdeok, and Pohang on the east coast, and Gunsan on the west coast. It was the first time in Korea that a local referendum was held on a single issue. Amazingly enough, this time it was all voluntary and no interveners had much say. Even more surprising, results came out when all the votes were counted. All four candidate localities showed overwhelming majority 'yes' votes; Gyeongju topped with 89.5 percent, Gunsan second with 84.4 percent, Yeongdeok third with 79.3 percent, and Pohang last with 67.5 percent. Finally, Gyeongju, in North Gyeongsang Province (right next to the Wolsong NPP site), won the competition. The long and turbulent road to a disposal site finally came to an end after twenty-some years spent with numerous civil riots and mistrust among people. The new Korea Radioactive-waste Management Corporation (KRMC) was created from KHNP to oversee the entire radwaste management issues from construction of the disposal site to actual shipping and storage of low- and medium-level wastes from all over the country. As of 2010, construction of the underground disposal facility at Gyeongju site is moving ahead at full speed, aiming to open for operation by 2013.

Perhaps the best known anti-nuclear advocate in Korea was from an NGO, Choi Yeol (president of Environment Foundation), who started organized protests against nuclear power in the 1980s. They claimed nuclear power was considered a form of pollution harmful to mankind, and thus must be eliminated. Numerous public debates held for pros and cons on nuclear power tried to convince public support on either side. The public generally tends to be swayed by emotional arguments raised by the anti-nuclear side while the scientific and safety arguments by the nuclear promoters often-times appeared in defense. However, decades of nuclear debates are finding a realism check that nuclear power is essential to the nation's economy and clean environment, provided the safety is guaranteed. Korean NPPs now accumulated more than one hundred reactor-years of operation without any major safety incidents. Climate change considerations and strong global trends for low-carbon economy is now shedding new light on environmental movements in Korea also, dampening the vocal and emotional anti-nuclear movements of the past two decades. Most vocal anti-nuclear NGOs are now shifting their missions toward new clean climate change movements, like the Environmental Foundation. KEPCO winning the UAE NPP contract in 2009 also played a pivotal role in changing the public perception dramatically on nuclear power to a more positive and proud endeavor, anticipating a new export industry to feed the nation's long-term economy, jobs for the young, and helping to curb global warming.

Lessons learned from the Korean radwaste disposal site issue were enormous. No one had anticipated the magnitude and rate of spreading public fear on anything nuclear could be so fearsome when it came down to the disposal site with local populations. Not only did it take twenty long years to finalize the site, but the ultimate cost to the tax payers skyrocketed with the incentive package to satisfy the increasing demands, plus the higher cost of deep geological cave method adopted. Ultimately, it could threaten the very basic economy of the nuclear power in the long term. Trying to press the high-level spent nuclear fuel together with the low/medium-level waste was a costly lesson. The national policy on spent fuel is still pending, with the final disposition of the spent fuel. Resource-poor countries like Korea would have no alternative but to recycle the residual energy contents out of the spent fuel once the international nonproliferation concerns were properly addressed; in particular, the North Korean nuclear issue needs to be resolved. Basic research on a novel approach to replace the existing chemical reprocessing of spent fuel, known as "pyro-processing," compared to the conventional Purex process, is under development in South Korea with international partners, as Generation IV reactor technology. The ultimate challenge for Korea is to prove the commercial competitiveness as well as the proliferation-resistant nature of the new technology.

Unfinished KEDO LWR

The Korean peninsula is geographically landmarked by two rivers separating it from China: the Yalu River ("Aprok-gang" in Korean), running westwards from Baekdu Mountain (the highest peak in the peninsula) to the Yellow Sea, and Tumen River ("Duman-gang" in Korean), running eastwards from the peak to the East Sea. Korea has been a unified country for over thirteen hundred years since the Unified Shilla period (seventh to tenth century), Koryo period (tenth to fourteenth century), Chosun period (fourteenth to twentieth century), and during the Japanese occupation period (1910-1945). Only during the last six decades since 1948 has the peninsula been divided into two: communist regime in the North (formally known as the Democratic People's Republic of Korea, DPRK) and market economy democracy in the South (the Republic of Korea, ROK). As such, two Koreas share the same history (except for the last sixty years), the same race, the same language (different dialects exist but are understandable by both Koreans), and most of all, a common desire to be reunified someday. Land area of the peninsula is about the size of the British main island; the northern area is about 1.5 times bigger than the south, but the population of the south at fifty million is about twice that of the north.

Both Koreas embarked on the reconstruction path following the devastating Korean War (1950-1953), which left both sides in shambles and poverty. Both Koreas had similar beginnings of the nuclear era in the 1950s by joining the IAEA and establishing the national nuclear research centers: Yongbyon (about one hundred kilometers north of Pyongyang) in the North, and Taenung in the suburb of Seoul initially, then moved to Daedeok in the South. Both Koreas' nuclear programs remained totally isolated from each other (even true today); however, they must have enjoyed the highest priorities from their governments. But the similarities came to an end in the mid-1970s, when the North took the nuclear weapons development path to further isolation and secrecy from the rest of the world. The first glimpse of their military intention was discovered by the IAEA in 1992, when their initial inventory report revealed serious inconsistencies, particularly in the amount and frequency of plutonium extractions from their graphite reactor of Magnox type at Yongbyon. Hans Blix, then the Director General of the IAEA, was among the first international delegates to visit the Yongbyon site in 1992. North Koreans indicated to Blix the willingness to convert their graphite reactors to LWR for the first time. However, the inconsistency issue evolved into the first crisis in 1993. North Korea unilaterally withdrew from the NPT as well as the IAEA membership, the first time ever by any country. The US and DPRK made a bilateral agreement (known as the "Agreed Framework") in 1994 to defuse all North Korea's weapons capabilities in the reactor, fuel, and reprocessing arenas. In return, two NPPs (light water reactor type which is far more proliferation-resistant) of 2,000 MWe size were to be built at

Kumho site in North Korea. A new international organization, Korean Peninsula (see "Nuclear Map of Korea" in page i.) Energy Development Organization (KEDO), was formed in 1995 to carry out the promise. In order to assure and overcome the mutual mistrust, the Agreed Framework was crafted into every reciprocating steps and milestones for both sides to abide. About ten years of difficult but intensive work was carried out to construct the KEDO LWRs in North Korea. The construction project came to a final stop when the second North Korean nuclear crisis erupted in 2002 with the suspected uranium enrichment program (UEP). Consequently, KEDO had to be dissolved in 2006 when the LWR construction was about 35 percent completed; over 1.5 billion dollars were spent, but it was unfinished. The world was astonished to discover the faint trace of xenon fission product gas from their first nuclear test in 2006, apparently leaked from an underground site at Punggye, Hamgyong Province, followed by the second test in 2009. The Six-Party Talks (USA, China, Russia, Japan, and two Koreas) under Chinese chairmanship has been on-again, off-again since the test in order to dissolve the North Korean weapons issue together with more permanent peace and security framework in the Korean Peninsula. (see "Nuclear Map of Korea" in page i.)

Above is the well publicized, brief summary of the North Korean nuclear puzzle story. Nearly twenty years have passed since the first North Korean nuclear crisis of 1992. It has been my personal passion and agony to follow up on the course of events from a professional interest. So many headlines were devoted to political, military, security, and diplomatic aspects of the issue while relatively little is known about the technical efforts mobilized in the process of realizing the KEDO LWR project. The following chapter will describe the LWR project from technical aspects as a logical extension of the NPP technology self-reliance program, which was happening simultaneously in South Korea. After all, over 1.5 billion dollars had been spent on the LWR project and 70 percent of it had to come out of the Korean tax payers' pockets. The story of the "unfinished project" needs to be told for the world to understand the ongoing situation that makes the Korean nuclear saga truly one-of-a-kind.

Which reactor for KEDO?

The basic policy line to supply the LWRs in return to eliminate the North Korean proliferation potentials was firming up among the US, DPRK, and ROK governments in 1993. The LWR issue was taken as a definite symbol of thawing tensions in the Korean Peninsula. Many rosy speculations flourished as to how and when North Korea would open up to the world, even to the extent that the US nuclear vendors like Westinghouse and CE were lobbying for their LWR plants to be marketed to North Korea. After much internal soul-searching, the South Korean government made a decision in 1993 to take up the leading position in the LWR project with the associated financial burdens, provided the Korean Standard Nuclear Plant was to be adopted. It was a bold decision to make from the South Korean side in 1993 since the first KSNP units, Ulchin 3&4, construction was just starting, with the Koreans taking the full responsibilities for the first time. The YGN construction under the US responsibilities was in its final phase. In fact, even the terminology of "Korean model NPP" caused serious internal debates among the Korean entities in that what we were building was the "US" design copy plants and should not be called "Korean standard" NPP. It was true that the design and the NPP technology were from the US origin, but also that Korean projects were the only fleet of new builds under full Korean responsibility. However, the very existence of large NPP construction under the national self-reliance program was not much known outside of the Korean nuclear entities, and thus the concept of the KSNP had to be foreign to the negotiating members of the LWR circle.

The role of one individual, Lee Byung-ryung, then the project manager of KAERI's Ulchin 3&4 NSSS system design, needs to be recognized. As a technical advisor's capacity to the Korean negotiating team (mostly from the diplomatic corps) for the LWR project, he made repeated pleas to the very existence and viability of the Ulchin 3&4 project to be adopted for the North Korean LWR. His unfaltering arguments subsequently provided enough confidence to the diplomats to be able to adopt the KSNP model NPP for the LWR project before the Agreed Framework was formally signed between the US and DPRK on October 21, 1994. It meant much more than building another NPP elsewhere, hoping to lay a bridging role to peace and stability in the peninsula towards reunification someday. A special Office of Planning for the LWR Project was created in Seoul under the Ministry of Unification to integrate inter-ministerial efforts, followed by the creation of international body KEDO in New York in 1995.

Given the LWRs were to be replicas of the latest KSNP being built in South Korea, who bore 70 percent of the total financing, the natural choice of the prime contractor for NPP contract was KEPCO together with the same team of subcontractors as in the Ulchin 3&4 project, plus some Japanese scope of supply as they bore 20 percent of the total cost. The main difference was the fact that KEPCO was no longer the utility owner and operator of the LWR but the turnkey NPP supplier to the North Korean Project Office through KEDO, and all project documentation was prepared in English to meet the international requirements. The construction site was cleared at Kumho in the South Hamgyong Province north of Hamheung (the second largest city in North Korea after Pyongyang) on the east coast. The Kumho site was originally surveyed by Russians in the 1980s for possible introduction of Russian VVER plants. Despite countless difficulties arising from working with the North Korean counterparts, design, manufacturing, and construction activities progressed sufficiently to pour the first concrete at Unit 1 on August 7, 2002, after receiving the construction permit from the North Korean regulatory authority. It was the first time ever the thousands of North and South Korean engineers and labor workers worked together side by side on a mammoth-scale construction project. Although both sides spoke the same Korean language, enormous gaps in mentality between the communistic way and liberal democratic way had to be overcome. After working together for several years, both sides began to narrow the mentality gaps and understand each other better. Long lead major equipment, including reactor vessels, steam generators, and the reactor vessel head assemblies, were being fabricated at the Changwon Doosan plant and Mitsubishi plant in Kobe, Japan.

When the Unit 1 containment building construction was at its peak in 2002, the second North Korean nuclear crisis erupted with the disclosure of the uranium enrichment program, which clearly violated the main premises of the Agreed Framework. Tension was escalating and the project was put to "invisible slow down" mode at first, then to "suspension," and finally to a grinding halt in December 2006 to terminate the KEDO LWR project. The only good news was the adoption of the extensive preservation and maintenance program at the Kumho site to protect and prevent from long-term erosion and corrosion damages to the civil and mechanical structures already built. A special storage building was erected at the Changwon Doosan plant to house the major equipment fabricated for long-term storage with the high level of preservation measures. Choi Han-kwon, having served in the special technical advisor's post at KEDO for the entire eleven years, lamented over the termination: "I feel confident that the unique experiences gained at the LWR project will play a pivotal role in the ultimate reunification process." Overall the construction progress rate was 34.5 percent at the time of the termination, after ten grueling years gone by, after some $1.5 billion was implemented. It appeared a sad end to the "unfinished" project.[1]

What could be the lessons learned from this "failed" project? Could anything positive be salvaged from the knowledge and experiences gained over the KEDO LWR project? This came much sooner than anyone anticipated. The first tangible outcome was realized just three years later when the KEPCO consortium was selected as the winner of the UAE contract in 2009, which is one of the overriding themes of this book. In fact, the KEDO project was the first NPP construction project in which KEPCO took the role of the prime contractor to supply the entire NPP under a turnkey project, instead of being at the receiving end of their own NPPs. The virtual customer was KEDO (not the DPRK), with multilateral composition in that every documentation and communication was conducted in English. In reality, the KEDO project was KEPCO's first export project outside of South Korea. The same KEDO project core members from KEPCO carried on to the UAE project bidding and implementation. No doubt the inside knowledge gained through KEDO played an invisible but crucial role in the UAE project.

How about the North-South relationship? Despite numerous military/political breakout episodes that happened (and are still happening) during the LWR project period, the rigid North Korean mentality was beginning to thaw from the engineering sincerity and integrity demonstrated by the South Korean participants. Initial North Korean fear of having the South Korean LWR in their own backyard that may turn into a "Trojan Horse" did fade away over the years. Mutual trust as a viable economic partner was realized at the Gaeseong Industrial Complex located just ten kilometers north of the DMZ border since 2006. Over one hundred South Korean small manufacturing firms invested their capital, equipment, and technology into the North Korean labor force, providing the North a taste of market economy. The four-party multilateral framework of KEDO (USA, Japan, EU and ROK) also laid a foundation for bigger Six-Party Talks since 2003, addressing a northeast Asian regional security issue. After all, the Six-Party Talks platform may be the only means to resolve the North Korean nuclear issue, together with the lasting peace and security in the region. I believe the "unfinished" LWR project may see its restart and completion someday only when the North Korean nonproliferation credentials are fully restored. Who knows? This day may come sooner than anyone dreamt of.

PART II: KNOW-HOWS AND KNOW-WHYS

In Part I of this book, focus was placed on the formation of the NPP technical self-reliance policy, which led to the onset of the YGN project, with the participation of KAERI in the construction project of the nation's tenth NPP in the 1980s. KAERI's role was to build up a technical capacity and organization in the areas of reactor system design and initial core design. The final decision to select Combustion Engineering over Westinghouse for the technology partner was the historic turning point in accelerating the Koreanization of nuclear power technology. However, it was only a part of the story. One needs to understand the history of nuclear power in Korea from the very first Kori Unit 1 to the Yonggwang Units 1&2, leading up to the YGN decision. Technical know-how's and NPP construction project management know-how's were gradually building up through the construction and operation of the first nine units. Momentum was building up for a dramatic change in the future fleet of Korean NPPs from the lessons learned the hard way.

CHAPTER 8:
THE FORMATIVE YEARS

"Atomic machine"

It is important to understand the role of Rhee Syngman (1875-1965), the founding first president of South Korea who shaped the new-born republic during his twelve years of presidential tenure (1948-1960). He is a legendary leader, a strong anti-communist, and a reformist of Japanese policies that had been imposed on Korea for nearly fifty years. Born into a noble family of the Yi dynasty, he was well educated from a traditional Confucius school in his early years, then later in the US for his college education, graduating from George Washington University, Harvard University, and finally receiving his PhD in history, politics, and economics in 1911 from Princeton University. His doctoral thesis from Princeton, *"Neutrality as Influenced by the United States,"* is a study of the history of neutral states since the eighteenth century and the unique role of the then-emerging power of the USA. By far, he was the best educated Korean politician of his time, far ahead of everyone else, with a clear vision for the world and the future of Korea. His prominent US education gave him a good understanding of a working democracy over communism when many nationalistic patriots were sympathetic to the communistic ideals, resulting in the splitting of the country into two Koreas in 1948.[1]

Being an independence movement leader during the Japanese occupation period (1910-1945), Rhee must have grasped the mighty power of the atomic bomb following the unconditional surrender of Japan in 1945. This notion of mighty power was further enhanced during the

Korean War (1950-1953) when the possible use of an atomic bomb against the invading Chinese army was seriously considered by the US.[2] At that time, peaceful use of atomic energy for electricity production was at its very infancy following President Eisenhower's "Atoms for Peace" speech at the UN General Assembly in 1953. Subsequently, the world's first commercial nuclear power plant, Calder Hall in the UK, went into commercial operation in 1956, and the International Atomic Energy Agency (IAEA) was created in 1957 as a special independent UN organization headquartered in Vienna, Austria. In fact, Soviet scientists had made an earlier move in demonstrating electricity generation from a nuclear reactor at Obninsk nuclear center (one hundred kilometers south of Moscow) in 1954. Historical records show that North Korean president Kim Il-sung visited the Obninsk reactor for its 1954 inauguration ceremony, providing a clue to the fact that even the North Korean leader had a keen interest in nuclear development from its infancy.

President Rhee is considered the 'atomic godfather' in South Korea for his personal vision and enthusiasm for the use of atomic energy, even for a dirt-poor Korea in the 1950s. His atomic vision was clearly focused on the peaceful use of nuclear energy to enhance production of electricity from the beginning for an emerging country like Korea. South Korea joined the newly created IAEA in 1957 as one of its founding charter member states while she was not even a member of the United Nations (the two Koreas joined the UN simultaneously in 1991, thirty-four years after joining the IAEA).

The only public remaining personal remark from President Rhee on nuclear energy is noteworthy. In his congratulatory speech at the construction ceremony for the Korea's very first research reactor, TRIGA Mark-II at KAERI on July 14, 1959 (less than one year before his forced exile from office in 1960 following the *April 19 Student Revolution*), President Rhee stated at the ceremony, "We must develop an excellent 'atomic machine' in the future," strongly implying the electricity from nuclear power. He deliberately chose and used this English term in an otherwise all-Korean-worded speech.

Walker Cisler

It took the visit of a well-known American engineer, Walker Lee Cisler (1897 – 1994), with President Rhee in Seoul in 1956 to prompt an initial move to nuclear power for peaceful use in Korea. Cisler (Cornell class of 1922, mechanical engineering) became famous with his highly successful mission to reconstruct electricity in Europe immediately after World War II under the Marshall Plan. He is also known as a tireless advocate of nuclear energy for electricity production. His first Korean connection came with providing electricity from generation ships anchored at Incheon and Busan harbors in 1948, when North Korea abruptly cut off electricity transmission to the South. He later became CEO of Detroit Edison and pioneered the first generation nuclear power plants for a commercial electric utility. He was also a founding member of the Atomic Industrial Forum (AIF) and National Academy of Engineering in the US.

When Cisler first met with President Rhee in 1956, only three years after the "Atoms for Peace" declaration by Eisenhower and in the same year of the first commercial nuclear power plant at Calder Hall in the UK being proven successful, he made a bold suggestion to the aging President Rhee (who was eighty-one then) to consider nuclear electricity generation in Korea. He somehow was able to convince the president of the magic of atoms by showing the "energy box," explaining the same weight of uranium would produce about three million times more energy than the same weight of coal. Simply put, he was trying to illustrate Einstein's theory of conservation of mass and energy, $E = mC^2$, where the energy produced is uranium mass loss (m) magnified by the speed of light (C) squared, as elementary Physics 101 courses do these days. In fact, this basic law of physics is the fundamental reason and attraction for nuclear energy; so much abundant energy from so little uranium substance is possible from quantum physics. When this astronomical energy is produced in a rapid and uncontrolled manner, it becomes the bomb; while the same energy is released in a super controlled slow manner, it can be harnessed for productive energy and electricity. Apparently impressed by the Cisler's "energy box" story, President Rhee asked two questions:

"What should we do first to make this energy possible?" and

"When do you think nuclear power will be possible in Korea?"

Cisler answered,

"You must establish an atomic energy office and atomic energy research center in the government and invest in training of nuclear scientists and engineers.

It may take about twenty years if all goes well."

The aging president took Cisler's advice almost literally with personal conviction, and history has proven Cisler right, for the very first commercial nuclear power plant, Kori Unit 1, went into operation in 1978, exactly twenty-two years after Cisler's prediction of the timing.

One argument that convinced President Rhee was the fact that Korea is an energy-resource-poor country (even more so in the south, as most of high-grade coal and hydro power dams are in North Korea). Cisler told the president, "All energy resources like coal are dug out of the ground, but nuclear energy is extracted mainly from the human brain. Countries like Korea have no energy resources except human resource, so start from technical manpower training." This is not an entirely correct statement since uranium is also dug out of the ground, as is coal. However, a significant difference is the amount of uranium is so much less than coal, but more highly trained technical manpower is needed for nuclear power production than for coal. The main attraction and lasting argument, even for today, for nuclear power over fossil fuel is that its dependence on uranium fuel is minimal while the dependence on high-tech human resources is enormous.

President Rhee wasted no time and made an executive order to create the Atomic Energy Bureau under the Ministry of Education in 1956. The first US/ROK Agreement for Cooperation Concerning Civil Uses of Atomic Energy was signed in the same year. A total of one hundred and twenty-seven young Korean scientists and engineers were sent to the USA for training under government fellowships, thanks to the bilateral agreement. Among the first group of fellows returned from US training was Yoon Se-won, who later became the first director of the Atomic Energy Bureau. Under his leadership, a nuclear law was drafted and was passed by the National Assembly in 1958, providing a firm basis for legal government commitment to the application of nuclear energy for peaceful use. The ROK joined the IAEA in 1957 as a charter founding member. It was a bold decision by President Rhee to take a personal interest in and commitment to nuclear matters while the country was still at poverty level (per capita income about seventy dollars) following the devastating Korean War. This laid a solid foundation of legal administrative infrastructure in Korea for the rapid deployment of nuclear power in the coming years. The remarkable stories describing the dawn of Korean nuclear history are well documented in a series of publications, "Founding history of Korean nuclear episodes," written by Park Ik-soo (1924-2006), who later served as the science advisor to President Kim Dae-jung.[3]

I was pleasantly surprised to discover yet another personal effort by the old President Rhee Syngman to understand exactly what the new nuclear energy could do for Korea in the 1950s. Apparently the president was briefed by several sources that nuclear energy could provide many wonders besides the bomb, and he wanted to find out exactly what the nuclear energy can do for Korea. He dispatched Chung Rak-eun in 1958, director of the Defense Research Institute in Seoul, to a four-month technical tour of the US, Europe, and the IAEA Conference in Geneva to obtain the latest information on nuclear power plant developments by visiting Calder Hall in the UK and Shippingport in the US. Chung, an electrical engineering graduate of Tokyo Institute of Technology during the colonial period, was the close confidant of President Rhee in the science and technology area. He left a travel report analyzing the NPP technologies of the 1950s. In summary, Chung reported to the president that the nuclear power for electricity production was the most appealing aspect of nuclear energy for a country with a keen shortage of electricity, like Korea. Chung left a personal memoir on his briefing to the president after the trip, and he was convinced that the old president was keen on the electricity production potential of the nuclear energy among all possible applications. This was amply proven that the new Atomic Energy Research Institute together with its first research reactor TRIGA was founded in 1959 only a few months after Chung's trip report to the president. What is most remarkable is the fact that the old President Rhee had the vision to lead Korea into nuclear power from the very beginning by collecting credible sources of information. Chung is probably the oldest living person in Korea today, at ninety-four, who was technically knowledgeable enough in NPP technology to lay a seed of hope for a country devastated by the Korean War.[4]

First feasibility study

An export-oriented five-year economic development program and preparation for the introduction of the very first NPP in Korea was well underway when the first oil crisis struck the world in 1973. A nuclear power investigation committee was first formed in 1966 at the Office of Atomic Energy (OAE), now under the newly established Ministry of Science & Technology (MOST), to initiate a feasibility study, including the candidate NPP sites. With technical assistance from the IAEA, Kori site north of Busan on the East Coast was selected to be the first NPP site among twenty-eight candidate sites. KEPCO was designated in 1968 to become the leading organization in planning, construction and operation of the new power systems on a strictly commercial basis. This was in direct contrast to the OAE's original plan to establish a separate government corporation solely devoted to the promotion of nuclear power, similar to the AECL of

Canada. Already by the late 1960s, South Korea was well poised to launch a major nuclear power program with the selection of the very first NPP site at Kori. The next order of business was to select the reactor type, timing, and the unit size.

The first indigenous technical feasibility study of the introduction of nuclear power was conducted in 1966 by the founding members of nuclear power from both AERI and KECO, headed by Lee Kwan, then the section head of the Reactor Engineering Department at AERI (he later served as the Minister of Science and Technology), with the following main conclusions:

- Korea could introduce its first NPP in the mid-1970s based on its economic development projections;
- The first NPP to be introduced could be the BWR and PWR types from the US of 500 MWe size;
- Capital cost of construction was estimated to be about one hundred million dollars and require the submittal of a loan application by 1968.[5]

One year later, an English feasibility study report was issued by a foreign consultant, Burns & Roe, an architect-engineering company in the US. It examined vast subject areas necessary for a newcomer country, similar to the present-day IAEA Milestone Document. The main conclusions were:

- Korea could afford to start its first NPP based on the economic development forecast.
- Three different power reactor types - PWR, BWR, and AGR - were proposed deployment in Korea without giving a strong preference. Two other types of NPPs were also considered: High Temperature Gas Cooled Reactor by the US General Atomics, and the CANDU reactor by the Canadian firm AECL. The B&R report concluded that these two should be eliminated based on the lack of commercial deployment for the HTGR and shortage of heavy water for the CANDU.
- The optimum size of the first NPP could be of 500 MWe class.

- Kori was the best candidate site for the first NPP.
  (The exact location of the reactor building was relocated to the outer rim at Kori from the inner bay location, as proposed by the OAE study).

PWR choice

Almost everyone in the nuclear industry today would not argue the fact that the Pressurized Water Reactor (PWR) is the single dominating reactor type in the world today. Over 60 percent of four hundred and thirty-seven operating NPP units are of PWR type. More impressive are the statistics on the new NPPs under construction. A whopping 84 percent of all (forty-seven units out of fifty-six total) new nuclear stations in the world are of PWR type, indicating the future global nuclear market trend. Five out of six commercially viable nuclear vendor companies today market only PWR reactors: Areva of France (EPR), Westinghouse of US/Japan (AP1000), Rosatom of Russia (VVER), Mitsubishi of Japan (APWR), and KEPCO of Korea (APR1400). When the Generation I NPPs were developed during the 1950s and 1960s, the UK and US led the pioneering new world of nuclear power with several reactor types. Britain was the first to commercialize with their Magnox-type gas-cooled reactor at Calder Hall, which later became the Advanced Gas-cooled Reactors (AGR). The American scientists from the Manhattan Project and nuclear submarine program went ahead with developing PWR types (by Westinghouse, Combustion Engineering, and Babcock & Wilcox), and BWR type (by General Electric). By the late 1960s, about an equal number of NPPs were being built and operated among PWR, BWR and AGR types, showing healthy competition among different technologies. In other words, no clear-cut dominating reactor type was apparent at that time. Subsequently, global nuclear power industries went through the boom and bust cycle of fortunes during the last half-century of the "dark age," and finally the renaissance. The PWR type must have endured the survival of the fittest through its safety and economics.

The selection of the nuclear reactor type for Korea's first Kori Unit 1 to be PWR, and not BWR nor AGR, was by no means an accident. One might deduce the fortunate outcome of the Korean NPP reactor types being all PWRs, except four units of CANDUs at Wolsong, to be sheer luck of a sort, judging from the international nuclear market at the time of the Kori Unit 1 decision in the late 1960s. But in-depth investigation of the decision-making process revealed otherwise. Three different types of power reactors - PWR, BWR, and AGR - were about equally successful commercially in the world market in those days, and the Korean nuclear community was split between the British AGR versus the American PWR (or BWR). Only a small number of power reactors were in commercial operation globally to be able to deduce any definite trends or performance statistics. Generally speaking, the US-trained nuclear specialists may have tended to favor the US reactors PWR (or BWR), while the British-educated specialists tended to favor the AGR-type in Korea. At this critical juncture, one KECO engineer had adequate nuclear power training and enough courage to speak out. It was Kim Chong-joo (1921-1993), who was trained at the Harwell Reactor School in 1959 and educated at the MIT engineering department, who made a lasting contribution.

Kim joined KECO as a young electrical engineer in 1952, when the country was still in shambles during the Korean War, moved up to the rank of vice president in 1964 in his forties, then became the executive vice president at KECO in 1972 (Kim later served as the vice president of KAERI, being the only person to have served at both KEPCO and KAERI as vice presidents). He commanded considerable respect among technocrats and Korean government senior officials alike. Recognizing the importance of selecting the right reactor type for the very first NPP in Korea, the Minister of Reconstruction (former body of Economic Planning Board, EPB), Tae Wan-sun, commissioned Kim in 1968 to produce a technical feasibility study report focusing on the reactor type. Kim recommended the US reactor types (preferably PWR) over the British one for the technical merits and reactor design characteristics. He was the only senior technical person in Korea who had sufficient knowledge about nuclear power at that time.[6] Unfortunately, a copy of this feasibility report is not traceable today. However, I was fortunate to discover a personal memoir of Kim Chong-joo published by his family after his death, based on his fifty years of diaries. He was a strict keeper of records with his personal feelings. It is interesting to note his comments in his diary of 1959 when he was sent for training at the UK Harwell Nuclear Lab and Calder Hall NPP to be trained as a nuclear power engineer. He wrote,

> "When I visited Calder Hall and asked the operator how many staff are working at the plant, they said 'no' because even that simple information was classified 'secret.' I began to wonder how this technology could be useful in commercial nuclear power if so much is shrouded in secrecy. I suspected their nuclear power technology had dual use of producing plutonium for weapons besides electricity. As it turned out, the British government banned all export of their AGR NPPs later, then subsequently shifted their domestic NPP types to PWR in the UK."

This one episode, plus his knowledge in the radioactivity contamination which is inherent in BWR design, must have given him the insight to put higher priority on PWR over AGR and BWR for Kori 1 recommendation.[7]

The first invitation-to-bid (ITB) document was sent out to two US PWR vendors (Westinghouse and Combustion Engineering), one US BWR vendor (GE), and one UK AGR vendor (British Nuclear Export) in 1968 as a turn-key contract with the condition to provide full financing. Soviet Union was also a PWR vendor at that time, but excluded from the ITB invitation as it was during the cold war period. Westinghouse was selected as the successful bidder together with the English Electric Company & George Whimpey of the UK. Although not successful, Combustion Engineering made their first attempt to enter the Korean nuclear market from the very beginning of Kori Unit 1. The selection of Westinghouse was based on the more favorable financing conditions rather than technical merit. No doubt the English version of the Burns& Roe report was an essential requirement for the loan application to the US EXIM Bank for the financing.[8] The first section chief of the Nuclear Power Section at the OAE, Kim Duck-seung, recollected those formative years:

> "If I compare what we planned in the 1960s with what we have achieved by 2010, I am humbled and overwhelmed by how closely our prediction of a half-century ago turned out to be the reality."[9]

Can do CANDU

Korea is the only country today with two distinct types of commercial nuclear power reactors besides China: the PWR-type light water reactor and CANDU-type heavy water reactor. While twenty-one units of PWRs are being operated, and more PWRs are being built and planned for future deployment at three nuclear sites (Kori, Ulchin, and Wolsong), CANDU stations are limited to the Wolsong site with four units in operation and no more. This is the outcome of the standardization policy in the early 1980s, when the nation's reactor strategy was determined – namely, to have PWRs as the primary reactor type, while the CANDU would be the complementary reactor type. In the late 1970s, there was a bigger plan for heavy water power reactors pursued at KAERI resulting in a joint technical feasibility study, "KC49 – Joint Canada Korea Study" by KAERI and AECL recommending deployment of four units of 900 MWe CANDUs with full technology transfer (refer to "**First two decades**" in Chapter 1 of Part I). This proposal was later scaled down to four units of 600 MWe CANDUs built at Wolsong during the 1990s since the nation's priority was set to establish technical self-reliance in PWR technology transfer with CE in 1986. Nevertheless, four CANDU stations at Wolsong represent the single largest CANDU station operating outside of Canada (other nations that imported CANDU reactors are: Argentina, Romania, China, India, and Pakistan).

Interest in heavy water reactors stemmed from basic neutron physics, utilizing the remaining fissile material from the PWR spent fuel (which still has more than 1 percent of U-235 fissile content), which could be reused as fresh fuel for the CANDU-type heavy water reactor. Due to its neutron characteristics, CANDUs are more adaptable to burn different types of fuels, including slightly enriched uranium fuel. A conceptual study on a tandem fuel cycle with PWR and CANDU types reactors was carried out at KAERI together with AECL in the 1980s to find a new area of fuel cycle research specially suited for Korea, with two types of power reactors in operation. It was called DUPIC, standing for "Direct Use of PWR fuel In CANDU," conceptually refabricating the PWR spent fuel without reprocessing in a proliferation-resistant manner into CANDU fresh fuel. DUPIC was well known in nuclear academic circles; however, the economics of such a novel fuel cycle prevented it from full-scale deployment.

As one may recall, the nation's second NPP was a CANDU-type Wolsong Unit 1 built in a full turnkey project with AECL of Canada in the early 1980s. KAERI was experimenting with the localization potential of several key reactor components, such as the end fittings (mouth piece at both ends of fuel channels in a CANDU reactor calandria vessel) and the nuclear fuel. Due to its simplicity in design and quantities required for continuous on-power refueling, the natural uranium fuel became the primary target of localization by KAERI. By 1987, KAERI was delivering a full core of reload fuels for Wolsong Unit 1 to KEPCO under a commercial supply contract (amounting to one hundred tons of fresh fuel per year). By this time, the entire South Korean nuclear entities were moving ahead with the YGN and Ulchin 3&4 projects with full-scale technology transfers from the US partners. When the Wolsong Units 2/3/4, 600 MWe CANDU replica plant of Wolsong Unit 1 construction project was started in 1990, KEPCO decided to follow the same contract scheme as the Ulchin 3&4 project. This meant the same EPGCC companies would take up the same responsibilities: KEPCO for total project management, KOPEC for architect-engineering, KAERI for reactor system design, and KHIC for major equipment supply with AECL as the subcontractor. A separate technology transfer contract on NSSS system design with reference to Wolsong Unit 3&4 between KAERI and AECL was signed in 1992, similar to the PWR technology transfer with CE. Experiences gained at Windsor in the PWR joint design activities were utilized in conducting the Wolsong project implementation, Unit 2 design, under AECL responsibility in Canada, gradually transferring to Korean responsibility in the subsequent units. One highlight of this program was to develop jointly the CANFLEX (CANDU Flexible) fuel between KAERI and AECL for utilizing slightly enriched uranium fuel for the advanced fuel cycle.[10]

CHAPTER 9: STANDARDIZATION

Once the national goal of NPP strategy and the self-reliance program was instituted and technology transfer partners selected, then came the real tasks of learning and implementing the technologies in full force. The construction start of the YGN in 1987 was such an epoch-making period not only for relevant Korean entities, but also for the global nuclear scene. The nuclear power programs in the US and Europe were about to enter into a dark age following the Chernobyl accident, with cancellations and moratoriums to stop and reduce all nuclear power projects for the next quarter of a century. South Korea may be the only exception (to a lesser degree in Japan and France). It turned out to be a true blessing for Korea to learn and master the NPP technologies through repeat construction projects. A staggering ten more KSNP copy plants (2 reactor units/plant) were built after the YGN: Ulchin 3&4, Yonggwang 5&6, Ulchin 5&6, Shin-Kori 1&2, and Shin-Wolsong 1&2. Invaluable experience was accumulated in mastering and improving design, manufacturing, and construction know-how's. Two more identical NPPs (reference plant Ulchin 3&4) were initiated at the Kumho site in North Korea in 2000 to implement the Agreed Framework, but subsequently were mothballed in 2006 after some 30 percent of the construction progress was stopped. (This topic is treated in Part I, Chapter 7: Lessons Learned Hard Way, **Unfinished KEDO LWR**).

Who says one can only be taught the know-hows, but not the know-whys? In a tightly controlled super complex of activities required to be performed for an NPP, and involving a large number of entities in Korea and the US, the first order of business was to learn how to copy it right from the technology partners in the US for commercially proven NPP technologies. This is basically learning to copy design and manufacturing processes employed on existing proven NPPs. Schedule constraints gave little or no time to ask: "Why do you do it the way you do?" In fact, the YGN system design was complex enough to scale down to 1,000 MWe power rating from the proven Palo Verde NPP's 1,300 MWe design. Only when Koreans learned how to replicate and verify it properly after several repeat projects did the time arrive to move on to improving and enhancing safety and economy of the NPPs, without sacrificing the merits of standardization.

Natural questions that confronted the Korean system designers were: "Why is the reactor design life-time limited to forty years?"; "Why is the fuel enrichment no more than 4.2 percent?"; "Why is the reactor vessel wall thickness twenty-five centimeters at maximum?"; "Why does the core protection system require a digital control package?" Such questions kept arising day after day with no clear answers readily available. Once you reach a certain level in the knowledge of the technologies, your inquisitiveness and curiosity are to seek the know-why's. These are the essential ingredients growing from the copy-technology approach into a mature ability that allows applying creative technology. In some cases large-scale verification testing was necessary to confirm crucial design parameters. One good example could be a large-scale integral effect high-temperature flow test loop at KAERI, called 'ATLAS,' to verify the indigenous simulation computer codes.

French example

The basic concept of standardization in any global industry requires a substantial development period in a competitive market economy. A good example can be found in the airline industry, where two main civilian aircraft designs dominate the world market today (Boeing and Airbus) with their standardized fleet of carriers. These two giants emerged after a larger number of aircraft vendors went through merger and integration processes during the rapid growth of the civil air travel industry in the last half-century. Today, the nuclear industry is now beginning to emerge out of the last thirty years of the 'dark ages' following the TMI and Chernobyl legacy in the 1980s. As such, leading nuclear vendor countries in America, Europe, and Japan did not have fully matured nuclear markets to apply standardized designs before the dark ages came. In the US, Westinghouse, GE, and CE were the major suppliers of NSSS, with different combination of many architect-engineering companies like Bechtel, Stone & Webster, Sargent & Lundy, Ebasco, and Gilbert. As a result, no two NPPs in the US had the same design and were generally customized to suit different utility's requirements. As such, the notion of building standardized plants was at most an idealistic dream.

The French had much better success with their standardized fleet of NPPs from CP1, CP2, to N4 series. The success was due largely because theirs was a sole utility company, EdF, which possessed the architect-engineering capability, and a sole nuclear vendor company, Framatome (which later became Areva). France today has the highest nuclear share of electricity generation in the world with nearly sixty NPPs delivering over 75 percent of the nation's electricity. This is credited to the French standardization policy of single utility and single supplier chain of companies, and to the national energy policies adopted by the French government supporting energy independence through nuclear power. Their domestic success story raised their international competitiveness to be able to export to South Africa's Koeberg NPP, China's Daya Bay NPP, and Korea's Ulchin 1&2 NPP in the 1980s and 1990s. The Korean nuclear infrastructure had many similarities with France (single utility, single supply chain of vendors) as well as strong governmental support. They were the natural candidate for bench marking in Korea when the standardization policy study was initiated in the early 1980s.

Comfort of dependence

Already nine NPPs were in operation or under construction in the early 1980s (Kori 1-4, Wolsong 1, Ulchin 1&2, and Yonggwang 1&2), all under turnkey or component approach contracts where the foreign

NSSS and turbine/generator vendor companies were the prime contractors. Except for Wolsong 1 (AECL of Canada), and Ulchin 1&2 (Framatome of France), all remaining six units were supplied by Westinghouse (2-loop plants at Kori 1&2, 3-loop plants at Kori 3&4, and Yonggwang 1&2). The first three units (Kori 1&2, Wolsong 1) were entirely under turnkey contracts where Westinghouse and AECL were responsible for the overall construction and performance warranty. Korean contents were kept at minimal levels in civil structures and operators training. Then came the next six units (Kori 3&4, Yonggwang 1&2, Ulchin 1&2) under the component approach. By then KEPCO was experienced enough to conduct the total project management directly using a component approach with several different prime contractors, including Westinghouse for NSSS, Bechtel for architect-engineering, and General Electric for turbine/generators. Kori 3&4 and Yonggwang 1&2 units were strategically selected to be copy plants to maximize the build-up of local technologies: architect-engineering by KOPEC and equipment manufacturing by KHIC, each as subcontractors to US prime contractors. KEPCO was gaining invaluable experiences in total project management to learn the know-how's in meeting the target schedule and budget without sacrificing quality.

Undisputedly, the US prime contractors Westinghouse and Bechtel duo was dominating the Korean nuclear market until the mid-1980s. They were the industry giants, known to be dictating their clients and aggressively lobbying the host governments. As a result, KEPCO/KHIC/KOPEC technocrats were groomed to be much more knowledgeable about the duo's technologies and tended to be biased against other vendors' technologies. Any notion of technical self-reliance was bound to be in line with the duo's technologies and their prerogatives. However, the Korean nuclear market dominance perhaps made the duo too arrogant. I was among many who felt the duo's reluctance in technology transfer had much to do with their arrogance. This eventually backfired when the ultimate winners were selected for the YGN project, and thus the future of the Korean nuclear market.

It is important to recognize, however, the contributions made during the Kori 3&4 and Yonggwang 1&2 projects in laying a solid foundation for the all-out technology self-reliance drive of the late 1980s. Without this preceding decade of preparation leading up to the YGN technology transfers, the Korean dream of developing an internationally competitive nuclear industry would have never been possible.[1]

The growing pains from the fact that four different reactor models by three different NSSS vendors from three different supplier countries were beginning to show a strain in all aspects of the Korean nuclear industry in the early 1980s, from the utility, KEPCO, to the regulatory ministry, MOST, which KAERI was actively supporting. In addition, an arrogance of technology "haves" against "have-nots" was keenly felt among those responsible for NPP operation and maintenance. For example, responses from overseas suppliers were poor in bringing in spare parts and services at reasonable costs and in a timely manner. To make it even more difficult, the dominating vendor company, Westinghouse, maintained an uncooperative and arrogant profile when it came to the transfer of the most significant technologies, resulting in creating a perpetual Westinghouse dependence. They were skilled in political lobbying to keep their dominance in Korea, but rather stingy about supporting local technical self-reliance, either refusing to provide crucial technology when requested or asking outrageous prices in return. Westinghouse concerns with confidentiality may have come from their nuclear Navy origin as a defense contractor. After all, Westinghouse's commercial PWR concept was a direct offspring of the US Navy submarine propulsion system, beginning with the Nautilus nuclear submarine in the 1950s. Their company mentality was much geared toward secrecy when it came to any sensitive technologies, as any defense contractor's should be. Dissatisfaction with Westinghouse and other foreign engineering companies, notably Bechtel, was brewing high among the Korean technical entities when the IBRD (who provided some loan arrangements for new NPP financing in the 1980s) conducted a study by Solomon Levy in 1982 on the Korean NPP program.

The minister of MER Suh Sang-cheol, made a bold decision to rectify the lessons learned, triggered by the Levy report, that led to the NPP standardization policy from the next units to be built starting with the YGN project. In short, the policy was adopted that all future NPPs in Korea would be based on a standardized design; domestic nuclear entities would be designated as prime contractors to reach technical self-reliance from the technology transfer licensing agreements; and safety and economic competitiveness would ultimately be raised. Thus, 1982 was the beginning of the local NPP technology drive and the standardization policy initiated from the Korean government after hard lessons learned during the preceding decade. Motivation to grow out of NPP technical slavery in core design, manufacturing, and operation/maintenance technologies was prevailing among the Korean entities. A revolutionary new approach to NPP construction was about to be unveiled with the onset of the next project, YGN (it was called the "Korean nuclear unit 11&12" project then, a misnomer carried away with the missing unit 4 saved for Wolsong 2).[2]

KSNP recommendations

The three-phase standardization study completed in 1986 recommended 1,000 MWe class PWR as the primary NPP model, with 600 MWe class PHWR as the complementary reactor model, as the candidate Korean Standard Nuclear Plant (KSNP) models. The study also produced a list of design improvements to be incorporated in the reference KSNP design. Korea already had nine NPP units in operation or under construction by then. The KSNP design was expected to overcome the current difficulties in multiple regulatory standards, equipment manufacturing cost escalation, non-compatibility of replacement parts, and overall drawbacks in indigenous localization efforts. Some outcomes of the study were also utilized in preparing the invitation-to-bid (ITB) documents for the Yonggwang 3&4 international tendering process. The study also identified four key NPP technology areas Korea needed to build up:

- Plant architect-engineering
- Reactor system design (including fuel and core design)
- Equipment component design and manufacturing (reactor and turbine/generator)
- Nuclear fuel fabrication

In hindsight, through the standardization study of the early 1980s, the importance of the nuclear reactor (NSSS) system design was recognized in defining the basic framework of NPP technology. Whoever controls the system design know-how must be the top-tier NPP technology holder, and who is most essential for being a true nuclear vendor.

Under strong MER leadership (with dedicated civil servants like Kim Se-jong, Director of Nuclear Power), an NPP standardization study group was formed with representatives from KEPCO, KOPEC, and KAERI, developing phase-I conclusions and recommendations. Kim, a graduate of Seoul National University in electrical engineering, was one of the principal architects in forming the government nuclear power policies in the 1980s. He later became the assistant minister of MOST in nuclear affairs, becoming the only civil servant having served both MER and MOST ministries. Shin Jae-in of KOPEC (a graduate of MIT nuclear engineering, who later served as KAERI president from 1994-1997) was the leading figure in the standardization study efforts to come up with the following recommendations:[3]

- 1,000 MWe-class PWR will be the principal NPP design model
- 600 MWe-class CANDU-PHWR will be the supplementary design model for Wolsong site

The preferential choice of a PWR reactor model over other well proven types like BWR and AGR reactor models in the early 1980s simply reflected the existing NPP reactor situation in Korea at that time (eight out of nine NPPs were PWR types), but in hindsight it was a true blessing for Korea and its export market potential. Today nearly 70 percent of global NPPs in operation and under construction are of one single type, the Pressurized Water Reactor, or PWR for short, dominating the world market.

It is important to note the birth of the Korean NPP standardization policy in the early 1980s in light of the following facts:
- Hard lessons were learned from three different reactor types, inhibiting local technical self-reliance;
- Dominating foreign suppliers' approach could be uncooperative in building local capacity, providing strong motivation to become technically independent;
- The continuing dynamic Korean economic growth, demanding enormous electricity generation, prompted adoption of a long-term nuclear power program in favor of expanding future NPP fleets;
- Recognition of the need to initiate a technical self-reliance drive with foreign technology transfer once and for all to build up domestic nuclear entities;
- Overall enhancement of NPP economics from construction project management capability, design standardization, operational capacity factor improvements, and safety regulatory licensing improvements.[4]

The standardization project was now blown into a full-scale project implementation stage to meet the project delivery schedule, as well as technology transfer from the US partners, CE, Sargent & Lundy and GE for the YGN project. It was to be a rough start, as the five-member Korean companies plus three US companies had never worked together before. In fact, Sargent & Lundy had never designed a containment building with a CE reactor inside before. Regular project review meetings (PRM) were called by KEPCO to oversee the progress. In addition, EPGCC workshops were conducted regularly from 1985 to 1995 in order to have in-depth discussions of technical details on a variety of issues arising from NPP construction under an indigenous approach. Never before had Korea witnessed the aggregation of technical minds coming together to reach the national goal of NPP construction technical self-reliance. Workshops were organized in rotation with central topics presented by the host institutes, normally held at the host site in Changwon, Daejeon, or company resort venues. A total of thirty-three

EPGCC Workshops were held in ten years, each had two-day presentation and discussions involving more than one hundred participants. KAERI played the secretariat's role in planning and documenting the workshop proceedings. Typical workshop topics included:
- NPP technical self-reliance progress and strategy
- NPP standardization progress and strategy
- Localization of NPP components and nuclear fuel
- Radwaste treatment and disposal
- Radiation protection enhancement plan
- NPP availability and capacity factor enhancement
- Nuclear safety and regulatory upgrades
- Public acceptance of nuclear power
- Advanced reactor design and new technologies
- Construction technology advancements

In addition to the five major Korean nuclear companies, nuclear maintenance, electrical, and natural gas energy companies joined the EPGCC in later years to bring more synergy among the indigenous energy technologies. This was the dawn of the Korean nuclear supply chain of companies, who learned to work together in developing higher quality and safety standards to be applied to the nuclear-grade components and systems. Through repeat construction projects of twelve KSNP plants in Korea over the last twenty years, every local company had a chance to qualify for nuclear work in a commercially competitive environment.

Once the government's basic nuclear power policy was set in 1982, the time had come to implement the new approach. The Electric Power Group Cooperation Council (EPGCC) of five local entities was designated with unique roles and responsibilities, as seen following.

KEPCO, the manager

KEPCO, the one and only electric utility company in Korea generating and distributing nearly 100 percent of all electricity (hydro, fossil, and nuclear) since 1948, had been relying heavily on foreign supply of generation facilities and technologies, mainly focusing on indigenous operation capabilities. As a reliable supplier of electricity in a rapidly growing domestic market, KEPCO was recognized as one of the blue chip companies by the Korean stock market with the highest AAA credit rating. The traditional conservatism of an electric utility company was a trademark of KEPCO, preferring the least risk when it comes to adopting local technologies. This was all about to change with the new role of the overall construction project management starting from the YGN project. It was the first project ever that KEPCO took up direct project management responsibility for and where local EPGCC companies took up the prime contractor's role together with foreign suppliers as subcontractors. Full-scope technology transfer was introduced for the first time among the prime/sub contractors under overall KEPCO supervision. In addition, it was the sole provider of financing for the construction and technology transfer contracts for the YGN project. In 2000, KEPCO was restructured into six generation companies covering different regions; however, all hydro and nuclear generation continued to come under a single company, Korea Hydro and Nuclear Power company (KHNP), a wholly owned subsidiary of KEPCO.

Since 1982, KEPCO played the central leading role in integrating all EPGCC companies toward the single mission of carrying out the NPP technical self-reliance drive, with its powerful and dedicated project management team. President Park Jung-ki, supported by Vice President Rieh Chong-hun, Lim Han-kwae (managing director of New Projects, responsible for the successful bidder selection of the YGN), Shim Chang-saeng (managing director of Nuclear Construction, responsible for the contract negotiation and implementation), and Park Yong-taek (better known as "YT", KEPCO project manager of the YGN project) were among the key personnel from KEPCO. YT Park was by far the most directly responsible for the success of the YGN project (he was a graduate of Seoul National University in mechanical engineering, and later served as the president of KOPEC). His best-known trademark quality was to make timely decisions and push strongly to proceed with high-quality products. Park reminisced his role as the PM: "I must be the luckiest man in the world to have been given the job of managing the most historic NPP project in Korea. I knew I had to push everyone to come to terms in working with each other, with no prior experiences working together. My job was to deliver timely decisions and make sure everyone kept up with every action item." The total YGN project cost amounted to over four billion dollars, including six hundred million dollars in technology transfer costs, the single largest construction project in Korean history at that time.

YT Park delivered the project on time, within budget, and with good quality, as he was the KEPCO PM for the entire project duration (he served as the Yonggwang site superintendant for 1994-1995 to personally witness the final commissioning phase of the YGN project).[5]

KOPEC, the designer

KOPEC is the sole nuclear architect-engineering company in Korea. It originated in 1975 from KABAR, which had been founded by KAERI and Burns & Roe, then was restructured into KNE in 1976 under KAERI. President Park Chung-hee authorized KNE in 1976 to be the sole nuclear architect-engineering company in Korea under KAERI management. This could be identified as the first official recognition of the nuclear national laboratory's direct participation in a commercial NPP sector on a project basis. KNE started intensive manpower training to work with global engineering companies, including Bechtel of the US and Belgatom of Belgium. KAERI management of KNE ended in 1982 when it was reorganized under the name KOPEC and became a wholly owned subsidiary of KEPCO, with Chung Kun-mo serving as its first president (he later served as the minister of MOST in the 1990s). While enjoying the position of sole source supplier of all nuclear architect-engineering roles in Korea, KOPEC had to compete in non-nuclear power engineering work to grow into an integrated global engineering company.

KOPEC was given the prime contractor's role of overall plant engineering for the first time in the YGN project, with its technology transfer partner Sargent & Lundy of Chicago. Yoo Joo-young was the first project manager, who led the plant design efforts. However, KOPEC's overall plant engineering work scope did not include the NSSS system design role at the beginning but had to wait until 1997 when the coveted role was transferred from KAERI (this story is described in Part I: Chapter 4).[6]

KHIC, the heavy weight

KHIC's participation in the YGN project as the prime contractor of NSSS and turbine/generator (T/G) provided a crucial turning point in its history. Chung In-yung (younger brother of Chung Ju-yung, founder of Hyundai Group) founded Hyundai International (which became KHIC in 1980) in the early 1970s with high ambition of becoming a world-class heavy industry company. The master plan of the Changwon plant was developed and built with the assistance of CE for casting, forging, and nuclear plant components, and of General Electric (GE) for the T/G plant. Financing was provided by the World Bank. Manufacturing license agreements were made with CE and GE respectively, in 1977 and 1976. It was one of the largest and most modern heavy industries built on a global scale at that time, comparable only to CE's Chattanooga plant and Framatome's Chalon plant. Once the Changwon plant was built in the early 1980s, the next big challenge was how to secure large-scale orders and necessary technical know-how to keep the plant operating in good standing.

The Korean government's decision to integrate and to streamline the power generation equipment manufacturers in 1980 was the definite life-saver for KHIC. KHIC was given exclusive rights to supply contracts for all domestic power generation equipment, including boilers, steam generators, and turbine/generators for fossil and nuclear plants. Prior to this decision, KHIC was competing with three other local heavy industries companies (Hyundai, Samsung, and Daewoo) for the domestic power generation market, which was not even big enough for a single vendor at that time. There was no real opportunity at that time to obtain export experience. KHIC was qualified to receive the ASME N-stamp for the first time in 1980 in preparation for the upcoming nuclear major equipment orders. KHIC gained valuable first nuclear vendor's experiences in producing the partial supply of steam generators and reactor vessels for Yonggwang 1&2 and Ulchin 1&2 as a subcontractor to Westinghouse and Framatome, respectively in the early 1980s.

The onset of the YGN project with a commitment to repeat copy plants for at least ten units was a necessary and sufficient opportunity for KHIC to bring its financial balance books into the black while delivering and accumulating critical manufacturing technical know-how and know-why's. Kwon Joong-gyu (later changed to Hong Young-soo) was the YGN project manager under vice president Chung Chung-woon, who was the mastermind behind all the technical cooperation schemes with CE. Large number of engineers were sent to the CE Chattanooga plant and GE Schenectady plant for education and training. Personal rapport and mentoring relationships were developed among the Korean and US counterparts to promote mutual confidence in design and manufacturing heavy components, such as reactor vessels, steam generators, and turbine/generators for the KSNP units. CE's open-minded corporate policy towards technology transfer, while recognizing the recipient would someday become their fierce competitor in the world market, turned out to be the main ingredient for success in a globally changing environment.[7]

KHIC remained the prime contractor for the overall NSSS and T/G supply from YGN and all subsequent projects, while KAERI retained the system design scope until 1997. At that time, the KAERI work scope was transferred to KOPEC. KOPEC continued to supply the system design work to KHIC. Specialty items like reactor coolant pumps (RCP), NSSS instrumentation and control (I&C) equipment, control element drive mechanism (CEDM), and reactor vessel internals (RVI) were excluded from domestic scope of supply and procured from CE at the beginning. However, KHIC began supplying RVIs and CEDMs beginning with Ulchin 3&4 project. RCPs and I&C equipment are being manufactured by KHIC for the 1400 MWe class now under construction. KHIC was finally privatized in 2000 through acquisition by Doosan Heavy Industries & Construction, and continues to supply major portions of new domestic and foreign NPP projects plus replacement heavy components such as steam generators. Today Doosan is by far the largest producer of desalination plants in the world, particularly well known to Middle Eastern customers like the UAE.

KAERI, the brain

KAERI's role in participating in the commercial NPP projects for a finite period (1985 to 1997) carries a special significance in the Koreanization story (much of how and why it happened is depicted in Part I of this book). No other country's national nuclear laboratory had direct involvements in its nation's commercial nuclear power program, as participants traditionally belong to the private sector nuclear industry and to electric utilities. Perhaps the only exception to this could be AECL of Canada, which played the turnkey nuclear vendor's role in NPP projects outside of Ontario Province (CANDU NPPs in Quebec, Argentina, Romania, and Korea), marketing and constructing AECL's own CANDU 600 models.

Since its founding in 1959 as the very first national nuclear research center, KAERI played the role of incubating new nuclear entities as the nation's nuclear program expanded during the past half-century. Some of these were commercial and others non-commercial, non-profit entities. The architect-engineering company, KOPEC (1982), and nuclear fuel fabrication company, KNF (1982), were created as daughter companies of KAERI and KEPCO, both as commercial entities. The nuclear safety regulatory body, KINS (1990), and nonproliferation safeguards center, KINAC (2006), became independent from KAERI to enhance transparency of regulatory control of nuclear energy in support of the government. In addition, the cancer research hospital in Seoul became KIRAMS (2002), independent from KAERI for medical and radiological research.

What makes the Koreanization story unique is the finite involvement of the national nuclear lab in highly competitive, commercial nuclear fuel and NPP construction projects based on technology transfer. It was a very controversial decision to designate KAERI to the YGN system design work at the beginning. The internal struggle between the drive for technical self-reliance and the controls of a commercial schedule control and warranty concerns for the plant construction was huge. It was necessary that high risk associated with unproven research-minded technical brains had to be taken while struggling to meet the project deadlines and budget control. Nevertheless, the risk-taking did turn out to be well worthwhile as the target of technical self-reliance goal of 95 percent was indeed met by 1995. The Yonggwang Unit 3 went into commercial operation on time, within budget, and surpassed all performance warranties beyond expectations. Being a national laboratory and not a commercial entity, KAERI's system design project team benefited immensely from basic KAERI infrastructure in saving costs and schedule. In hindsight, KAERI's involvement for twelve years (1985-1997) of power projects was the right time at the right place by the right people under the inspiring right leadership of Han Pil-soon. I was the first project manager for the first six years, then the second half had Han Ki-in, who served until the end of the YGN project and realized the technical self-reliance goal of 95 percent.

KNFC, the core

One important member of EPGCC was the nuclear fuel fabricator KNFC. It was created in 1982 with a single mission to fabricate all nuclear fuels for domestic NPPs, as a daughter company of KEPCO and KAERI. KEPCO supplies all uranium feed material in natural or enriched form, and KNFC conducts uranium conversion and fabrication service to the final fuel assemblies for its sole customer, KEPCO. It is fully equipped with initial and reload core design software capability, as well as fuel parts and assembly hardware technologies. Reload core technologies were already established with KWU of Germany, and later with Westinghouse, for its NPPs Kori 1-4 and Yonggwang 1-2 units.

During the YGN project, KNFC was the prime contractor to KEPCO on nuclear fuel supply, while KAERI had a separate initial core design team with CE, like the system design team, but under subcontract to KNFC. Kim Jae-poong was the project manager for KNFC throughout the entire project period.

Hyundai, the builder

Last but not the least in the EPGCC was the group of local construction companies headed by Hyundai Construction. It was founded by its legendary leader, Chung Ju-yung (1915-2001), creator of, at one time, South Korea's largest multinational conglomerate *chaebol*. Hyundai Construction was responsible for site construction of almost all domestic NPP projects from the very beginning, Kori Unit 1. Their specialty was to implement the site civil/structural construction, mechanical, and instrumentation managements suitable to meet the nuclear-grade quality control and quality assurance. The most difficult task was to secure qualified welders for the nuclear-grade site welding processes. A number of innovative approaches in reducing the construction period while securing quality standards, such as more modular construction approach, was introduced.

While the foreign NPP constructors nearly disappeared during the last twenty-five years due to the nuclear Dark Age, Korean constructors such as Hyundai accumulated invaluable experience unparalleled anywhere. This was why the construction area was not included in the technology transfer package with the US partners in the YGN project. Cha In-whan was the project manager for Hyundai during the entire period. The current Korean President Lee Myung-bak is personally very knowledgeable in nuclear industry in general and NPP construction in particular, as he was a Hyundai Construction man moving up to the CEO post when Hyundai was the prime contractor for many domestic NPP construction projects. Korea was fortunate to have President Lee as the head of state, who had personal knowledge of the NPP industry and was able to make direct contributions when the historic UAE contract was finalized in 2009.

CHAPTER 10:
KOREAN PILGRIMS

Design technology independence!

It was the day before my forty-second birthday, December 12, 1986. My wife, Lucia, was not so happy that I had come home late as usual - too late for a special dinner she had planned. There was a good reason for coming home late on my birthday-eve, as we had to prepare a special farewell party for the first group of the KAERI technical team (called "Group A") to be sent to Windsor, Connecticut, for the first time. Altogether, forty-six Group A team members were given one-way tickets to New York (Korean Air KE 081 from Seoul to JFK). If I only knew the significance of this flight, we should have spread the team onto several different flights (one accident could have been a disaster - a loss of Korean nuclear technology independence all together). We had to make a very special occasion to remember this day. All Group A members and KAERI senior managements were gathered at the newly built VIP hall on the second floor of the main building at the Daedeok KAERI campus. I opened the meeting with a brief report to the audience. Contents of my report in a nutshell were as follows: only seventeen months before, the Power Reactor System Division was created at KAERI to start the NPP system design work with thirty-three members. It had grown to be over one hundred with internal transfers and external recruits. They were selected for their nuclear academic or engineering background as well as English language capability. An intensive domestic training program was undertaken to prepare the new team for mobilization. IAEA experts were brought in to Daejeon to give classroom lectures, followed by on-the-job training at the Kori NPP site to familiarize them with the nuclear steam supply system at a real operating NPP. Many of them had never seen a live nuclear power plant before, let alone knew how to design one. Motivation was heightened as they began to realize they were to become the first Korean engineers to be able to master the know-how of the power reactor system design and become an integral part of the ambitious Korean nuclear industry.

After fierce competitive bidding, finally Combustion Engineering had been announced as the successful bidder for the YGN supply and technology transfer only ten weeks before, in September 1986. Still, the final contracts with the US partners (CE, Sargent & Lundy, and GE) were not yet signed; hundreds of detailed contract terms and conditions were under intense negotiations. More than anything else, we had no time to wait for the final contract. As a result, we had to make an early start with some risk-taking once we knew who would be our mentor. Everyone was working very hard to sign the contracts soon, but the negotiations were getting tough as the Korean sides were making more demanding terms. One of the demands from the KAERI side was to start the US training of the Joint Design team from January 1, 1987, even before the contract was signed. CE accepted this proposal, fully knowing the risks involved as hundreds of KAERI engineers would be sent to Windsor, USA. The decision was made jointly by CE and KAERI managements to move into the on-the-job training at Windsor, Connecticut, where the CE head office was located, even before the main supply contract signing. Full members of the KAERI system design team would be relocated to Windsor for the in-situ training preceding the actual Joint System Design work to begin as soon as the contracts were signed. Time had come to dispatch the first group of the KAERI Joint Design team to Windsor.

Han Pil-soon made a moving speech at the farewell luncheon. "You are all modern-day Pilgrims, about to discover the new world of nuclear power technology in the US, where it was born. Each of you is given a one-way ticket to New York. We will fully back you up and expect you to succeed in your mission, but do not expect to come home if you don't succeed."

"Our mission may appear impossible to you as it was to Admiral Yi Soon-shin in the naval battle against the Japanese, we can win the battle only if all are willing to die for the mission". Legendary Yi beat the invading Hideyoshi's navy at all 23 battles in 23 encounters with paramount shortages of ships and crews in the 16th century. It was a chilling message. Everyone realized the gravity of the missions on their shoulders. Han raised a framed calligraphy plate high up in the air. It read: "必設計技術自立," written in Chinese brush hand-writing, meaning, "*We Shall Have Design Technology Independence.*" It was about two meters high, seventy centimeters wide, sealed in a wooden frame with a glass cover. No one knew who wrote the calligraphy or how one could prepare powerful art-work in such a short notice, but the message was loud and clear, in line with Han's speech. Their mission was to master the NPP reactor system design know-how's, and they were about to go to the place of origin of the technology in the US. It was a solemn reminder to all in the room of the goal set for each one. At the end of the farewell event, everyone joined the three rounds of "hurrahs!" at the top of their lungs, shouting, "필 설계 기술 자립!, 필 설계 기술 자립!, 필 설계 기술 자립!" to promise, "*We shall reach design technology independence!*" three times.

Today, the same calligraphy message is carved on a stone monument at the KAERI campus in Daejeon. The monument was erected in 1997, exactly eleven years after the shouts of "hurrah," to commemorate the successful YGN construction and achievements of reactor system and initial core design technical self-reliance. By then, they had achieved the ability to successfully replicate the System 80 design and had contributed to the successful completion of the project as major team members, enough reasons to celebrate and leave a permanent monument at the birth place of design technology self-reliance.

Expedition support

It was by far the single largest overseas expedition of KAERI staff in its entire fifty-year history - never before done, and never after repeated. A total of about two hundred KAERI staff members were dispatched to

Windsor for a long-term duration of more than twelve months each. It spanned over the eleven-year period of 1986 to 1997 for the sole mission of the YGN Joint Design, together with technology transfer. Many of them were stationed at Windsor for periods of two years. The total accumulated KAERI manpower that stayed in CE Windsor amounted to nearly three hundred man-years (not counting numerous short visits to Windsor for meetings). Most of the KAERI engineers were in their late twenties and thirties, with good technical academic backgrounds and reasonable command of English. Most of them had engineering MS or PhD degrees, many of them from well-known US universities such as MIT, University of Michigan, Penn State, RPI, and North Carolina State University. What they were lacking was the hands-on design experience of nuclear reactor systems and perhaps fluency in English. All of them were male staff, and most of them were newlyweds. At that time, overseas travel outside of Korea (even for a week) was a rare event and considered a privilege by others, especially to be able to take along the family for a long period. Many of them had never taken an airplane ride before, let alone for an extended period of several years to live in the US together with their families. For a young newlywed to take along his bride to the US for a couple of years (fully supported by the company) must have been very self-motivating for any Korean engineer.

Getting a passport and visa in those days was a big deal. ROK government had just recently relaxed passport requirements of permitting husbands and wives to travel overseas together, one liberalization process as the 1988 Seoul Olympic was approaching. KAERI management made a strategic decision to dispatch all project team members with their families from the beginning to implement the YGN project and technology transfer, as spelled out in the Joint System Design contract with CE. Issuance of individual passports and US special work visas, together with airline tickets for the whole family (SEL to JFK), were fully supported by the administrative staff at Daejeon as well as in-situ support at Windsor. A special post of KAERI Windsor Office site manager was created to head up the large group of expatriates living in a strange country for the first time. The following senior KAERI staff served in the site manager's role during the peak period at Windsor, supporting over two hundred Korean expatriate staff and their families:

> Kim Jin-soo (1987-1989), nuclear engineering graduate from SNU, manager of safety analysis
> Lee Ik-hwan (1989-1991), nuclear engineering graduate from Hanyang University, recently recruited from Hyundai Construction Co. (he later served as the president of KNFC)
> Shin Hyun-kook (1991-1993), electrical engineering graduate from Hanyang University, manager of Instrument & Control

One must remember the YGN contractual conditions imposed on both CE and Korean entities at that time. CE was ultimately held responsible for the performance and delivery warranty of the NPP to go into commercial operation by 1995, although they were contractually the subcontractor to KHIC and KAERI. In other words, CE was responsible for keeping the schedule and the quality (thus the performance) of the NSSS supply for the entire YGN project. They had to control the NSSS project progress schedule as if they were the prime contractor. In reality, both KHIC and KAERI had to follow CE's lead in every aspect of the project. CE's project management team at Windsor, headed by Jim Veirs, the Korean Project Director, Mike Barnoski, the Project Manager, and James Crawford, the NSSS Engineering Director was keenly aware of this fact, more so than anybody else. The sudden influx of a large number of Korean system design engineers under their disposal must have created a huge challenge and logistics problem at their Windsor headquarters office.

The KAERI site manager's role was exactly to work as the counterpart to CE management in dealing with everyday working and living conditions for the expatriates. Kim Jin-soo, the very first KAERI site manager to arrive in Windsor on April, 1987, and Lee Ik-hwan, from 1989, had the initial teething problems of settling down in a new environment supporting all Korean staff, family members, and frequent Korean visitors. A wide range of issues, from finding housing, school accommodations for children in elementary and middle schools, buying used cars, and finding hospitals for medical treatments, etc., etc., had to be addressed by the site manager. Most fortunate was the CE staff's willingness to help out the Koreans any way they could, plus the overall friendly and welcoming atmosphere at the Windsor community towards the Korean newcomers.

During the peak period of 1987 to 1990, well over two hundred Korean staff and families were working and living at Windsor, creating a small colony of Koreans at an otherwise mainly Caucasian north eastern small town. More than seventy Korean babies were recorded born at Windsor, as many of the engineers were newlyweds in their most reproductive life cycle. They were indeed "Korean Pilgrims" to have landed at Windsor for a finite duration with a specific mission. Resemblance with the original Pilgrims from England in the seventeenth century was that they had no way of returning home unless they succeeded in their missions. Subsequently, the KAERI Windsor Office was changed to KOPEC Windsor Office from 1997, reduced to only a few staff to function as a KOPEC/CE liaison office until 2007. Today, only Doosan has its liaison office at the Windsor Westinghouse office.

In retrospect, it seemed like a miracle that the Korean mission to learn the know-what's and know-how's in the power reactor and fuel design technologies was successfully completed by 1995, as the newly built Yonggwang Unit 3 went into commercial operation on time. The technical self-reliance goal was also achieved to allow the next replica plant, Ulchin 3&4, to have the Korean entities now taking the design lead with US support (exactly the opposite of US/Korean roles from the YGN), and taking the full performance warranty as the prime contractors.

Not a single Korean expatriate failed to achieve his given mission or dropped out of the project during and after his Windsor stay for the entire three hundred man-years of expedition throughout the period.

About 15 percent of the entire Windsor-trained staff subsequently left the nuclear system designer's profession due to natural turn over to go to other jobs after they returned to KAERI and completed their duty and obligations to the nuclear program; others immigrated to the US and Canada. Most unfortunately, two colleagues from the original Windsor team, Yang Seung-yeong, manager of quality assurance, and Han Jae-bok, section head of instrumentation & control, passed away of illness after their return from Windsor. The Korean Pilgrims played the most crucial role in the history of NPP technical self-reliance and all of their names will be remembered.

One significant observation is made here. It was KAERI, the national nuclear laboratory founded quarter of a century before, that was fully in charge of the Windsor expedition as a prime contractor to KEPCO/KHIC during this period. No doubt they had the best-educated nuclear-related engineers in the country. That was the main rationale for designating the system designer's role to the national laboratory in the first place instead of to a private company. What was not appreciated at that time, however, was the benefit of using the national lab's management in carrying out the unprecedented overseas expedition to accomplish the technology transfer. Until then, KAERI was accustomed to send scientists abroad (mainly to the US and UK) for extended training with very limited foreign-aid funding. This time was different, as the fixed price project funding was sufficient and fully secured. KAERI was prompt and responsive in setting up new sets of administrative procedures and regulations in dealing with the Windsor expedition. In addition to all technical matters to be digested, new administrative support, including travel allowances, family support stipends, reporting and accounting procedures, passport/visa issuances, medical insurance coverage while at Windsor, and transfer of foreign currency, was established for the first time.

Unlike any private industry practice, KAERI implemented fully 100 percent of its project funding from the YGN project (about one hundred million dollars for the whole scope of NSSS system design and initial core design, and corresponding technology transfer) to cover its direct expenditures during the 1987-1995 period. No concept of indirect costs (such as profits, company overheads, etc.) was even considered, as it was not allowed in the accounting practices of the national laboratory. The industry practice of putting aside about 30 percent for overhead cost was unheard of and unthinkable under KAERI management at that time. In retrospect, it was one of the main blessings for project success by allowing KAERI to recruit the best available manpower and put them into action in the overseas expedition fully covered by the project cost. Private industry would have been far more stingy and limited in implementation in a fixed-price contract like the YGN project. The fact that not a single drop-out occurred during the whole Windsor expedition testifies clearly to the soundness of the management over the project duration. In summary, the mission was completed to the satisfaction of everyone involved, including the host, CE, and everyone returned home safely after their assignments and contributed in each one's specialty area.

Remembering Windsor

I had to make the trip. It was much more than a sentimental journey back to old Windsor, Connecticut, in the north eastern corner of the US after twenty years of absence. I was a frequent visitor to Windsor during the peak period (1985-1991), three or four times a year. Several hundred KAERI engineers were sent there for training and joint design work at the beginning, for which I was directly responsible as the KAERI project manager. I had to find out from the old memories of the CE counterparts how they viewed the entire Korean project and how the integration of US/Korean joint design team worked out from the US perspective. After all, writing only about the Korean side of the story would be only half-complete or biased, I thought.

In mid-October, 2010, Windsor was already at the peak of autumn colors with brilliant maple reds and birch yellows. I was confident that my oldest CE colleague, Jim Veirs (then the one and only Korean project director, from Vanderbilt University in Tennessee with a degree in electrical engineering, and a nuclear Navy graduate), made all the arrangements for me to interview. I was fortunate to meet the "old pros" from CE: James Crawford (then the NSSS Engineering Director), Regis Matzie (then the Reactor Engineering Manager, graduate of Stanford with a PhD in nuclear engineering, and nuclear Navy officer, who later served as the vice president of Westinghouse Nuclear Research), Tom Natan (then the one and only Engineering Manager for the YGN project, born in Austria, graduate of UC Berkeley in mechanical engineering), and Mark Crump (then the Technology Transfer Manager for the YGN, graduate of the University of Michigan with a PhD in nuclear engineering). They were all happy to recall with me in person the unique experiences we had together nearly a quarter of a century ago. I could feel their enthusiasm in reminiscing about the old stories as they saw some value in my writing the book on the subject.

The first thing I wanted to ask Veirs and Matzie was on the UAE project:

> "What was your first impression when you just heard the news last December that the Koreans won the UAE contract?" I was expecting their surprised feedback.
>
> "Well, it didn't surprise me a bit, as we knew the Koreans had the best chance."
> "UAE was most keen on the delivery date of 2017 and wanted a strong utility operator's participation, as they were seeking joint operation after the plants are built. In that respect, Koreans had the best chance over French or American bids. KEPCO being the only nuclear plant constructor and operator among three bidders, was the natural choice, not a surprise choice, in addition to the proven safety and economy."
> "Now the UAE deal is setting a new model of NPP export projects for many developing countries seeking nuclear electricity with operations commitment from the supplier."

"Bottom line: the Korean price tag must have been lower than the competitors' as the risk overhead multiplier factors were well assured by the Korean government."

I thought there was a striking analogy between the selection of CE in 1986 for the YGN project and the Koreans in 2009 for the UAE project; both winners were the least-known vendors at that time with no previous overseas NPP market experiences. I thought perhaps the market economy of selecting the most reliable and economic NPP played well in the UAE, as it did in Korea twenty-three years ago, or "technology over politics" was the case in point for both countries.

Some stories, like the history of Windsor and the CE Nuclear, did not hold much of my interest then, but now I find them interesting to recognize their significance as I am getting older and am in a process of restructuring the quarter-of-a-century-old history that took place on both sides of the Pacific.

Windsor was founded in 1633, the very first Pilgrim settlement in the state of Connecticut, only twenty years after the original Pilgrim Fathers landed at Plymouth on the Mayflower. It was the Puritans who came from England seeking liberty and religious freedom, starting the Congregational Church from the old Anglican Church upon arriving in the New World. Several tribes of indigenous Indians occupied the Connecticut Valley; the most aggressive ones, like Pequots and Mohegans, had been at war with each other in many battles, leaving a minor tribe, like the Podunks, to invite the white-men to settle in the Windsor area as a mediating force. The history of Windsor recalls that the Podunk tribe chief selected the English Pilgrims over the more powerful Dutch settlers, who were a dominating merchant power in the New England region in the early seventeenth century. A peaceful deal was made between the Podunk tribe and the Plymouth Pilgrims to make the first settlement at Windsor. Today a historic landmark boulder can be found on the campus of Loomis-Chafee in Windsor, testifying to the Pilgrims very first settlement in 1633. Descendants of these Pilgrims are the backbone of today's Windsor community.[1]

History may be repeating itself, I thought, some three hundred and fifty years later. The Koreans selected CE over Westinghouse in the Yonggwang competitive bidding. KAERI played the crucial role in the bid evaluation. We were most impressed by the neighborhood honesty and sincerity of commitment to transferring CE technology, compared to Westinghouse's arrogant and commercial attitudes. With a bit of imagination, CE played the Pilgrims, Westinghouse the Dutch, while Koreans played the Podunk Indian tribe's role by choosing CE over Westinghouse. The congenial environment of the Windsor community in general, and the open-minded CE corporate culture, helped to win Korean friendship in the latter part of the twentieth century. I reckon Windsor had its roots firmly imbedded from the early Pilgrim days, helping Koreans to reach the formidable goal. This is how the Korean engineers at Windsor came to be known as the "Korean Pilgrims."

Descendants from the early Pilgrims may constitute the majority of the present-day Windsor community population, backbone to the northeastern New England states in the US. Their character can be summarized as the most liberal democratic state with the nation's highest education and intellectual standards. Some of the world's most renowned, top ranking universities like Harvard, Yale and MIT are all in the vicinity of Windsor, providing intellectual grounds of cooperation. I have no doubt that the CE's corporate culture of open-mindedness and the Windsor community culture of open-heartedness provided invisible but immensely fertile soil for the Korean Pilgrims in achieving their goal of technical self-reliance.

History of CE Nuclear

CE Nuclear entered the nuclear field as a contractor to the US Navy, first as supplier of reactor pressure vessels from its Chattanooga facilities and as the designer of a submarine reactor and fuel manufacturer located in Windsor, Connecticut. The commercial nuclear business has its roots in the acquisition in 1955 of General Nuclear Engineering Company (GNEC), a small engineering design company that was started by Dr. Walter Zinn. Zinn had been the director of Argonne National Laboratory and later founded GNEC in order to become more active in commercial nuclear power. His reputation as a pioneer in nuclear physics had been made as a member of Enrico Fermi's team that designed and built the Chicago Pile located at the University of Chicago's Stag Field. It was here for the first time that the sustained nuclear chain reaction was demonstrated. Zinn and Harold Lichtenberger were present at this historic event. Lichtenberger, who joined Zinn and later became a key CE executive, held the axe that was to be used to cut the rope to insert the poison rod should the experiment in nuclear fission dictate such action at the Chicago Pile.

Along with Lichtenberger, Zinn, both of whom it could be said were "present at the creation," assembled an experienced team at GNEC, including John West who would be Zinn's successor at CE after his retirement. After being acquired, GNEC brought its experienced design team to join with CE's existing capability and pioneered CE's first commercial pressurized water reactor design. This enabled CE to begin competing in the 1960s with Westinghouse and General Electric. After several successful deployments of Generation I PWRs at the Palisades and other utility sites in the US, CE established its brand name with the "System 80 Generation II PWR" (meaning 'a reactor for the 1980s') NSSS design to market during the first nuclear boom period of the 1970s. Frank Bevilacqua, vice president of Engineering at CE, could be called the "father of System 80" design development.

CE was generally credited with a superior engineering and design, evidenced by the fact that the megawatt yield of its nuclear reactors was typically about 10 percent higher than that of comparable Westinghouse plants. The basis for this increase in efficiency was from a computer-based system of core monitoring called the Core Operating Limit Supervisory System (COLSS), and Core Protection Calculator (CPC), which leveraged large numbers of in-core neutron detectors and a patented algorithm to allow higher-power densities. As their company title, "Engineering," implies, they had superior engineering ahead of everyone else in the business, as exemplified by the COLSS-CPC, fully digital control of the thermal power output from the nuclear reactor.

CE had sixteen System 80 units under firm order to be built when the TMI accident broke out in 1979. Three units of Palo Verde NPP were the only System 80 NPPs built and operated, while all others were cancelled. CE also competed in the international markets during this period, and was selected as the successful bidder for Taiwan's Lungmen project in 1982, but subsequently the project was cancelled. CE lost to Westinghouse in the Kori 3&4 project bidding. Their first connection to the Korean nuclear project was to supply the reactor vessel for Kori Unit 1 as a subcontractor to Westinghouse in the early 1970s. The golden period for CE in the new nuclear plant business had arrived in 1986, when they won the YGN contract. It was the beginning of a quarter-of-a-century relationship between CE and the Korean nuclear industry. From the late 1980s on, Korean projects were the only new plant business for CE while the rest of the industry suffered near global nuclear darkness following the Chernobyl accident.

The corporate governance structure of CE went through several dramatic changes since the Korean projects began. The first restructuring came in 1990 when Asea Brown Boveri (ABB), a European power generation giant, took over CE to form ABB-CE. However, the nuclear portion of CE stayed pretty much intact to continue on with the Korean projects; only the name changed. A much bigger change came in 2000 when Westinghouse, traditional archrival in the nuclear sector, took over control of CE Nuclear following ABB-CE's financial crisis, due in part to the outcome of law suits resulting from a large number of fossil power plants' asbestos health effects. Although CE Nuclear was a profitable operation in the 1990s due to the Korean projects, it became a "gem" used to salvage the parent company, ABB, by its sale to Westinghouse, which was then owned by British Nuclear Fuel Limited (BNFL). In

2006, Toshiba of Japan, traditionally a BWR nuclear vendor company, acquired Westinghouse in a strategic move to globalize nuclear vendor companies in preparation for the coming age of nuclear renaissance. From what can be observed from the outside so far, the Toshiba/Westinghouse merger appears to be financial in nature, while leaving the nuclear technical and marketing aspects in Westinghouse hands.

Under Westinghouse, not only the name "CE" disappeared for good, but also the flagship brand *System 80* technology was dropped altogether in lieu of AP1000 technology, developed by Westinghouse. Only through the Korean projects, the *System 80* technology endured, was improved, and survived. I thought that "old die-hard pros" from CE, like James Crawford and Tom Natan, must have special feeling about the APR1400. It is an improved version of CE's System 80+, developed by the Koreans and winning of the UAE contract, which all started from the YGN in the 1980s. In a recent interview at Windsor, I asked Crawford and Natan,

> "What do the Korean projects mean to you?"
> "The YGN project was the most successful, best job I ever had. We had great people from Korea to work with. We had greatest respect for YT Park (the YGN project manager from KEPCO). He was a pain at that time but surely got the job done by making the right decisions on time, by forcing us to do the best job we could possibly deliver."

I said,

> "CE was paying a price for its success in Korea, as Westinghouse bought out CE."
> "Hmm, it's an interesting way to look at, but, true, they needed us to win the China market for sure, but they had no interests in the System 80 technology, only the Korean technology transfer experiences," Natan lamented.

It was an honest expression of pride and agony coming from the best mentors with whom we worked together. One can say the legacy of CE technology is now solely in the hands of their once-mentee and licensee, the Koreans, who have made it competitive in the world market.[2]

I was much saddened to hear of the untimely deaths of Bob Newman, then the vice president of CE Nuclear Systems, Herb Cahn, then the director of International Business Development, and Vince Krecicki, then the resident manager at KHIC Changwon. We will remember them for their personal and enthusiastic commitments to the Korean project, and always with a unique sense of humor.

The Korean projects were the landmark achievements of CE Nuclear's history since the mid-1980s. To realize these achievements, CE had to go through a huge learning experience as well to cope with hundreds of young Korean engineers coming to their backyard to learn their technology, as they had had no previous experience on foreign nuclear projects before. Not only was the company's policy of full-scope technology transfer to the Koreans clear to everyone from the CE side, my feeling was the New England Pilgrims' mentality of sincerity and congeniality toward the engineers helped to set the right atmosphere in working together with the Koreans coming to Windsor.

At the early stage, CE hired a consultant Karl Markovitz, owner of Korea Strategy Associates, to come to Windsor for CE staff training on Korean culture and mentality. Markovitz, who held a PhD in history as well as a Harvard MBA, had spent many years in Korea and spoke fluent Korean. Through these sessions, as well as others that were conducted from time to time over the twenty-five year period, CE staff had a glimpse of Korean history and mentality, what made them happy, what made them angry, etc. etc. Jim Veirs, the project director during the entire Korean project, could be the one single person most responsible for resolving multitudes of technical and nontechnical issues and problems from the CE side. He made over one hundred trips to Korea over the past twenty-five plus years, more than any one I knew, gaining confidence and respect from Koreans. His wife, Linda, a retired librarian from a local town library, whispered in my ear when I was departing from Windsor on a recent visit:

> "I knew all along you Koreans and my husband did many good things for Korea and Combustion Engineering, but he never talked much about it at home. I hope your book will tell the whole story for my children to know what their father did." I simply told her,
>
> "I will do my best, but… wish me luck."

The same situation is true with me also, as I never had enough energy or time left to tell my nuclear work story when I came home at nights. What better incentive do I need to finish this book, in all honesty? Many of us in the business could share the same desire to be able to tell his own family and children what we did in a more objective and proud way. All the ingredients for success were ready at Windsor: the best mentors with the most advanced technology and the willingness to teach and share everything they had, and the best mentees fully motivated with high educations and willingness to outperform anyone's expectations.

CHAPTER 11:
THE JOINT DESIGN

The original concept of the Joint Design was a brainchild of Han Pil-soon, as crystallized in his inauguration speech as KAERI president on April 9, 1984. He said, "The only way for a developing country to reach nuclear power technical self-reliance is to take part in the commercial NPP construction projects in the areas of KAERI's expertise. We have no money, time, and experienced manpower to bring it up from scratch." That was the beginning of a long road to bring KAERI to take responsibilities in the NPP reactor systems and initial core design technologies by taking a prime contractor's role in the YGN project.

The first Joint Design concept was put into practice in 1985, when the PWR reload fuel localization project was started with KWU of Germany. Already seventy-plus KAERI fuel engineers were sent to KWU fuel design center in Erlangen, Germany, to conduct reload core analysis together with the German counterparts in a joint design under the technology transfer contract. All design tools were transferred to KAERI before they returned to Korea by 1987. The first reload core fuels designed by KWU and KAERI, then manufactured by KNFC in Daedeok, were being loaded into Kori units. This successful experience of Joint Design in the reload fuel projects gave a strong impetus for a bigger and more complicated NSSS reactor systems design project to be conducted in a similar fashion. Now the stage was set to expand the fuel experience to the entire NPP reactor systems with different actors.

Preparation

Immediately after CE was selected as the successful bidder of the YGN and NSSS partner company in September of 1986, the Power Reactor Systems Division at KAERI was realigned to be the same as the CE Windsor organization. We knew right away the most efficient way to absorb the system design technology would be to emulate the mentor-to-be's technical infrastructure as much as possible to prepare for the upcoming Joint Design missions on a one-to-one approach in five technical disciplines: reactor engineering (RE), fluid systems engineering (FSE), mechanical design and analysis (MDA), instrument and control (I&C), and safety analysis (SA). The manpower pool within KAERI had reached over eighty when the very first team of forty-six (Group A) was sent to Windsor at the end of 1986, having grown from thirty-three members just about a year before. Already word had gotten around in and out of KAERI that the most ambitious nuclear project in history was about to begin. We were assured and thus fortunate to be able to hire new recruits outside of the government manpower authorization limits, since the entire manpower cost was to be supported by the Yonggwang project cost once the contracts were signed. The new PRSD team was composed of five technical groups of highly educated engineers with graduate degrees (about 50 percent with MSs, 30 percent with PhDs, and 20 percent with BSs) in engineering disciplines, mainly nuclear, mechanical, electrical, chemical, and physics backgrounds. The initial PRSD organization management team in 1986 was as follows, with the respective CE NSSS engineering managers:

	KAERI	CE
RE	Kuh Jung-eui (Ha Young-joon)	Regis Matzie
FSE	Lee Byung-ryung (Han Gyu-seong)	Joe Longo
MDA	Sohn Gap-heon (Park Sung-ho)	Dan Peck
I&C	Hahm Chang-shik (Han Jae-bok)	Bill Gill
SA	Kim Dong-soo (Koh Byung-ryung)	Fred Carpentino

(KAERI names in parenthesis are the initial team leaders at Windsor)

In addition, a high level of English proficiency was required to ensure the capability to be able to absorb details about the coming mission.

The Joint Design

Conversational English training courses were held at KAERI with native US speakers. With all their academic credentials, the KAERI team may have appeared to be well- or even over-qualified; however, we had a definite shortage of (in fact, zero) experienced manpower with NPP systems design backgrounds. No one with such qualifications existed in Korea as of 1986. We were all young, well educated, and well motivated, but had no practical experiences to speak of. The only way to overcome this shortage was to put the team through a series of rigorous training courses in combinations of class-room training (CRT), on-the-job training (OJT), and on-the-job participation (OJP), later called "Joint Design" or "Joint System Design," plus field trips to NPPs and component manufacturing sites.

One crucial six-week CRT course was organized by the IAEA Technical Cooperation project specially tailored for the KAERI system design engineers at the KAERI campus in October 1986. A brief Memorandum of Understanding for the system design advanced training was signed between CE and KAERI only two weeks after the successful bidder announcement. Everyone was eager and willing to take the risk, to get on with the training even though the main contract negotiations had yet to start. We were able to bring in most of the CE engineering managers or supervisors as the lecturers. CE management put high priority in sending their best technical managers, all of them for the first time, to Korea. Regis Matzie (reactor engineering), Dan Peck (mechanical design and analysis), and Bill Gill (instrument and control) were among the first group of CE managers who came to Korea. During and after the formal CRT hours, they were enthusiastic about the first opportunity to get to know each of their Korean counterparts besides delivering the technical materials. Matzie recalls his first visit to KAERI:

> "I was surprised to meet so many young PhDs, but Kuh Jung-eui and his team made me feel right at home from the very first day. We went out for dinner with the whole RE group, got to know them personally, and I was impressed with their engineering enthusiasm. I felt they had the right DNA in mental capacity to master the system design know-how's in

no time. Kuh's team appeared more academic than practical, but it suited fine with my area of reactor engineering."[1]

Obviously this kind of initial building of rapport had lasting effects as they moved to Windsor for the Joint Design in the coming months. Mental comradeship was forming between the mentors and mentees among the engineers, supervisors, and managers from both sides, overcoming the cultural and geographical barriers. Perhaps the degree of comradery between two organizations would differ from team to team depending on the personal character of each of the managers, but I felt overall comfortable enough to judge that we were in for a long haul, and we had the right teachers and right mentors from the very beginning.

We took the CE lecturers to Gyeongju, the ancient capital of the Shilla dynasty, after the first CRT, which turned out to be effective in understanding Korean culture and nurturing the team-building. (They saw the original Cheonmachong Shilla golden crown at the Gyeongju National Museum then.) As a symbol of lasting cooperation and friendship, KAERI presented the Shilla golden crown to CE in 1988 (fortunately, this one symbolic piece still proudly occupies the main entrance hall of the Westinghouse Windsor plant today). This small gesture of cooperation made over two decades ago is worth revisiting today. It exemplifies how the original dream of technical self-reliance on the Korean side became a reality with CE as the strong partner and mentor, as engraved on a small plaque attached to the crown, reading:

> "This exact replica of the ancient Shilla Gold Crown (Cheonmachong, circa 5th century A.D.) is presented to Combustion Engineering, Inc. on the occasion of the KAERI Windsor Site Office opening ceremony held on the twenty-ninth of January 1988. This symbolizes long lasting partnership between CE and KAERI through this Joint Design and Technology Transfer during the implementation of Yonggwang 3 & 4 Project. CE's contribution in helping to establish indigenous nuclear

power technologies is recognized for the energy independence in the Republic of Korea."

Technology transfer, Korean style

It took a relatively short time before the "Korean Pilgrims" started to be integrated into the CE Windsor NSSS engineering departments. After a brief two-week orientation CRT given to the newly arrived Group A Korean engineers to Windsor in January 1987, each one was assigned to his respective NSSS engineering department for the following four months of OJT. This was the crucial period for the KAERI team to prove themselves worthy of conducting the NSSS system design activities for the first time, while fully being integrated into the existing CE infrastructure.

At the end of the four-month OJT program, every Korean engineer was evaluated and declared fit for service without a single failure. Despite the English language deficiencies, every team member survived the initial tests of functioning as a design engineer to initiate, to review, and to obtain supervisors' approval on a mock-up design activity. The first informal progress report from Lee Byung-ryung ('BR Lee,' the first engineering manager of the KAERI team at Windsor) was sent to President Han via a seventeen pages of hand-written facsimile message at the end of March 1987.

Lee was expressing his satisfaction on the progress of the OJT of each team member in preparation for the real Joint Design to start as soon as the formal Yonggwang contract was signed. Detailed curriculum of subtasks, objective, design description, output products, interface requirements, and schedule of design activities for each member were being implemented as agreed. CE supervisors were evaluating KAERI members' knowledge on the technical subject matter as well as communication skills. Lee's overall conclusion in the fax report was confidence in the system design technical areas *per se*; however, he made a recommendation to improve the project management capabilities in planning, scheduling, manpower loading, interface coordination, and, most of all, quality control and quality assurance. These were the areas the KAERI management was not well prepared for at the beginning. In fact, we did not know what it would take to integrate vast system design activities across different disciplines in reaching the major milestones in the project.

Lee's recommendation was well taken once the contract was duly signed, and the full scope Joint System Design activities commenced in May 1987. Thanks to the hard work of several industry-minded KAERI project management teams at Windsor, the need to establish an Engineering Logic Network was recognized by the KAERI management. They were able to acquire sufficient know-how from CE internal practices on how best to set up the KAERI specific system design integration processes, coupled with the schedule and cost control. A separate KAERI Windsor Site Office was created with the special mission to build up the project management capacities, including schedule control, quality control, and document control. Lee made an important contribution during the initial period of KAERI's settlement into the *bona fide* system design engineer's role at Windsor. All members were asked to come into work thirty minutes earlier and stay two hours after the normal working hours, even on Saturdays, for internal seminars. No one dared to complain as there was so much to learn and absorb. Lee provided leadership to promote useful means to better organize one's work activities, to learn about other members' work activities, as well as to improve English communication skills, including how to raise questions to CE mentors. In hindsight, his major role as the KAERI engineering manager in Windsor at the early stage is to be recognized in building team spirit. In the 1990s, Lee made one more major contribution to the Korean nuclear scene while serving as the project manager for the Ulchin 3&4 system design at KAERI, in selecting the KSNP model for the KEDO project with Ulchin 3&4 as the reference plant. (This topic is covered in Chapter 7 of this book.) Lee also served as the Yuseong District Office superintendant, an elected public official post in Daejeon city, administering for Daedeok Science Town in the 2000s.

An additional wave of KAERI design engineers arrived in Windsor in 1987 and 1988 in Group B (thirteen) and Group C (twenty-two) to join the Joint System Design activities following the similar CRT and OJT training as Group A. Altogether, over two hundred KAERI engineers were dispatched during the project life cycle, which peaked to one hundred and twenty in 1988, accumulating over three hundred man-years of expatriate experience at Windsor. Beginning in 1989, KAERI manpower receded gradually as they completed the Joint Design assignments and returned home (the Initial Core Design team was the first to return) to build on the project design center at Daedeok.[2]

Integration with the mentor

Designing and building a nuclear power plant is a complex task involving thousands of highly trained, dedicated professionals in several dozen different organizations, mostly made up of engineers in many disciplines. Among the most difficult, thus challenging, would be the system design of the reactor system, or in the industry jargon called Nuclear Steam Supply System ('N triple S'). An added dimension to an already complex task on our hands was to integrate organizations from two countries from opposite sides of the Pacific, to start the design process in the US first and then to move the entire design center to Korea to complete the second half under the so-called "Transition Plan."

Reactor system design activities are categorized in three different levels. Level 1 design is the "conceptual design," involving several hundred independent activities to identify the skeleton of the reactor system for the intended power level, the process system, and the safety system. Level 2 design involves roughly eighteen hundred separate design activities, called "basic design," enough to issue the Preliminary Safety Analysis Report (PSAR) and design specifications for the long-lead items such as the reactor vessel and steam generators. Level 3, called "detail design," typically involves twenty thousand design activities to generate all the documents and drawings for construction, procurements, and final commissioning start-up tests. When a utility orders a new NPP, typically the reactor vendor has a certified basic design from the national regulatory authority, and the detail design has to be redone from the reference plant design to suit the specific site and upgrade items as necessary.

Difficulty in the job assignments for KAERI members arose during the first year of Joint System Design at Windsor in 1987. It had to occur as both parties were bound to have conflicting views. KAERI's team was determined to learn the core technology in the shortest time possible while stationed at Windsor (each one was given a two-year term) so they wanted to get to the most challenging design activities as possible. In addition, KAERI was in the customer's position to CE, although they shared a mentor/mentee relationship. We made sure in the Joint System Design contract that KAERI would provide 50 percent of the overall design activities in volume without knowing exactly which 50 percent would be their share.

On the other hand, CE was given the overall responsibility and contractual warranty obligations should they fail to meet the delivery deadlines or performance warranty. Each of KAERI member's capabilities and communication skills was still not fully proven for CE to be comfortable, although they all had passed the minimum qualification testing after the initial four months OJT at Windsor. This mutually uncomfortable situation was more acute in the fluid systems engineering than others, mainly because of the nature of design activities being more repetitive and less innovative, and perhaps due to the personalities involved. It could be boring work for someone with a PhD degree and at a relative young age. In the area of instrumentation and control, often-times the design activities crossed boundaries between the system design and component design due to its very nature, thereby adding to the complexity. (Although BR Lee, the engineering manager, had some problems with job assignments, he was up against some tough-minded CE managers and had to make stands that were not always popular. All in all, he did a good job in a situation that was unprecedented.) As the years passed by, however, these initial teething problems did get resolved with satisfaction to both sides. I could confirm the facts through many interviews

The Joint Design

among many from both sides during the writing of this book. Of course, the final outcome of the technology transfer without a single drop-out among over two hundred KAERI expatriate engineers speaks for itself.[3]

Invariably the system design activities involved producing documents, be it design specifications and design data for procurement or manufacturing, or interface documents to other entities like the architect engineers, or calculation sheets and drawings. Almost all of these documents are considered company proprietary and become part of the company's intellectual property. Each document is approved and signed off by the writer (cognizant engineer), an independent reviewer, and an approval authority from the technical supervisors. The integrated team of the Joint System Design members learned to produce literally thousands of these documents in a highly professional manner. The system of document control, known as the Nuclear Engineering Documentation System (NEDS), originated by CE during the previous Palo Verde project, was employed. This document control system added order to the management of these documents and proved quite beneficial to all parties during the course of the project. The system allowed virtually free access to all technical proprietary documents to Koreans at any time. The document control system was indicative of CE's open and transparent corporate mentality.

This corporate attitude was the main difference from its competitor when we selected CE over Westinghouse in the first place, and was found to be faithfully practiced during the course of the project. I could safely say that CE's attitude was to give Koreans everything they had, every 'know-how' they knew, without any reservations. At one time in the early period, CE Windsor had their internal discussion on the subject of to what level CE's proprietary technology should be open to the Koreans. Vice President Bob Newman made this point clear to everyone:

> "Remember, this opportunity is once in our lifetime. If we don't succeed in transferring all we can to the Koreans, then our technology would be of no use anymore, and we might as well throw it into the Connecticut River."[4]

This one simple anecdote represented the entire corporate culture of CE on the technology transfer. It was a true blessing for Korea in reaching its goal of self-reliance. KAERI was able to establish its own document control system administered by the Design Document Control Center (DDCC) at Daedeok with a fully functioning electronic database accessible from any authorized desk-top PC in the intranet.

Transition Plan

The original plan to relocate the system design center physically from CE to KAERI was imbedded in the Joint System Design and the Technology Transfer contract documents, stipulating that we shall reach a concrete transition plan within eighteen months after the contract. It was one of the most important tasks given to me as the project manager of the YGN System Design to reach a mutually agreeable plan with my counterpart, James Crawford, director of NSSS Engineering at CE. After many meetings and deliberations, we were able to sign a document called the "Transition Plan" on May 1, 1988. It was to implement the physical relocation of manpower and design tools, including computer codes, design documents, and supporting infrastructures. Within a year's time, the Yonggwang reactor system design center would be fully operational at KAERI's Daedeok campus without interrupting the overall project schedule. This plan would also involve relocation of CE supervisory personnel to live in Daejeon for a longer period (a total of fifty man-years of CE supervisory staff was allocated for Daedeok assignments) with their families, plus gradual return of all KAERI expatriate engineers back to their home base. It was an ambitious plan to achieve. The overall project milestone of PSAR submittal to the Korean regulatory body and serious licensing discussions were beginning in mid-1988. Heavy components like the reactor vessel and steam generators were being casted and forged at the KHIC Changwon plant based on our system design specifications, and the construction activities were moving into high gear by Hyundai Construction Company at Yonggwang site. There was no question in anyone's mind that

The Joint Design

the center of gravity for the project was moving to Korea. We had to hurry back to be closer to the real actions as a consequence of the system design output.

The first CE technical staff who came to Daejeon for an extended period was Robert Kirkman from Windsor in January 1988 to install and verify the computer codes at the KAERI computer center. His job was to hand-carry the source codes together with the binary codes for the system design. Over one hundred and ten codes (CE-owned codes, third-party-owned codes, plus some commercial codes) were installed at KAERI, followed by thorough verification checks.

By mid-1988, all design codes were running flawlessly in time for the KAERI design team's return home. Together with the documents that had already been delivered to KAERI beginning in 1987, the Design Center was well prepared to support the continuation of the system design in Daejeon. Over the course of the project, forty-six hundred proprietary documents and drawings were physically relocated to or developed by the Design Document Control Center.

Kirkman returned back to Windsor after a six-month stay in Daejeon. His working and living experience in Korea alleviated much worry from many CE staff who had some apprehension of coming to live in Korea according to the Transition Plan. In fact, his first hand working and living experience as a first-timer in a strange foreign country with a totally different culture turned out to be quite positive. His experience demonstrated to CE that comfortable living accommodations were readily available in a welcoming Korean community where local people were anxious to assist in any way. Undoubtedly it was more of a culture shock for CE staff to live in Daejeon than for KAERI staff to live in Windsor since no CE engineers spoke Korean, while the KAERI engineers were reasonably proficient in English.

The initial batch of KAERI engineers began returning home by the end of 1988. On April 18, 1989, CE Daedeok Office was formally opened with Jack Pucak as its first site manager. Pucak's role was mainly to liaise design activities between Windsor and Daedeok. Tom Natan, the engineering manager for the YGN, and Dan Peck, manager of mechanical design and analysis, came to stay three to four months in Daedeok to expedite the in-situ decision-making.

With more than two years of Joint System Design experiences at Windsor and the gradual return of experienced design engineers to Korea, and with the design infrastructures being ready with codes and documents in full operational mode, we felt reasonably comfortable to move the design center to Daedeok in 1989. Integration of the system design support team at Daedeok with the newly returning KAERI engineers from Windsor posed less of a problem than expected, as a continuous inflow and outflow program to and from Windsor was in place. Every move was in line with the Transition Plan. The first major task at the new design center was to finalize the Final Safety Analysis Report (FSAR) reflecting all the design upgrades and specifics as the plant was being built. Over eleven hundred design activities were scheduled to be performed at Daedeok, with KAERI engineers in the lead and CE supervisory functions as needed. Start-up test guidelines and regulatory updates were among the highest-priority items. A real motivation for the system engineers was the opportunity to eye-witness the manufacturing and construction of the components and systems they designed for verification and trouble-shooting any anomalies with implementing the design. Frequent visits to the KHIC Changwon plant, where the major components like the reactor vessels and steam generators were being fabricated, and to the construction site at Yonggwang became a routine part of the design verification activities. KAERI Yonggwang Site Office was formally opened in 1992 to expedite the start-up support activities.

Warranty conditions

The YGN contract was indeed one-of-a-kind in many aspects, requiring the foreign subcontractors like CE and S&L to take full warranty responsibilities while being the subcontractors to the Korean prime contractors: KHIC, KAERI, KOPEC, and KNFC. Since the Korean prime contractors had no proven past credentials of delivering the whole NPP at that time, KEPCO had to make a special provision to induce CE and S&L to assume the warranty responsibilities by making the subcontracts into tripartite agreements including the KEPCO signature. This measure was intended to provide additional protection in case the prime contractors failed to deliver on their roles and responsibilities. However, as it turned out, this additional protection was not necessary as the Korean prime contractors all managed to meet their obligations. CE had to take the largest share of warranty obligations as the NSSS technology holder. The scope of their warranties included:

- The thermal output performance at the steam generator outlet
- Delivery schedule
- System design accuracy
- Interface data correctness
- Regulatory licensing (from the Korean authority)
- The fuel cycle costs

It was clear from the beginning that the technology holder, CE, was the only qualified entity to assume the above warrantee conditions. These conditions included clear formulas for assessing the financial penalties in case any of the warrantees were not met. The most important of them all was the thermal output performance warranty. This warranty included a financial penalty that could accrue in the millions of dollars should the thermal output be less than the warranted designed value. It was a measure of the success of the project that the Yonggwang Project demonstrated all warranted conditions, and there were no contractual claims against any of the parties following start-up and full power demonstration testing. The plant was unconditionally accepted for commercial operation by KEPCO on March 30, 1995.

The true sign of technical self-reliance must come with the ability to take up the full warranty conditions, including the financial burden. The Korean local prime contractors (which were identical to the YGN case) had to wait for the second NPP project, Ulchin 3&4, which was signed in 1991, to assume all the warranty obligations for the first time. This was the main reason Ulchin 3&4 NPPs were the first KSNP units to be called "Korean Standardized Nuclear Plant" (KSNP).

Manpower build-up

One may recall the original system design technical staff was only thirty-three when the Power Reactor Systems Design division was initiated at KAERI in 1985. In the next ten years' time, these core members witnessed a dramatic growth to about four hundred technical staff responsible for the reactor system design of multiple NPP construction projects ongoing simultaneously: YGN, Ulchin 3&4, Yonggwang 5&6, KEDO project for North Korea, and Ulchin 5&6. By the 1995 timeframe, the entire KAERI project operation was at its peak with five project divisions (Yonggwang, Ulchin, Wolsong, LWR fuel, and KEDO) supported by the Design Division in matrix organization. As the NPP projects had about a two-year time interval between new projects, a continuous flux of manpower from one project to another, and from design division to project division (or vice versa), was necessary to maintain the proper level of manpower loading for each project at any given time. A project management team was dedicated to each project for interface control, schedule, cost, and quality control, in addition to managing the site offices away from the Daedeok head office (Windsor, Yonggwang, Ulchin, and Wolsong construction sites).

The system design technology was central to and rested in the Design Division composed of five technical departments: Reactor Engineering, Fluid Systems Engineering, Mechanical Design, Instrumentation & Control, and Safety Analysis. The YGN occupied the largest manpower share (nine hundred and eighty-eight man-years, at least) to cope with the technology transfer simultaneously with the joint system design, which started at Windsor first then transferred to Daedeok in the middle of the project. While the Joint System Design (JSD) manpower was implementing the design activities according to the project construction schedule, additional System Design Support manpower was allocated to follow the technology transfer in digesting technical know-how. The following table illustrates how much overall KAERI system design manpower actually participated during the entire ten year project period of 1987 to 1996 for the YGN project alone:

JSD	Windsor	133 man-years
	Daedeok	164
SD Support	Daedeok	494
Project Management		155
Quality Assurance		44
Training		58
Start-up support		15
Total		1,063 man-years

CE mobilized about three hundred man-years of effort for the first project, which dropped to minimal after several repeat projects. More JSD manpower was needed at Daedeok after the Transition Plan. About half of the overall manpower was dedicated to the SD Support activities at Daedeok to conduct activities including:

- Mock-up design and repeat design on technically significant activities
- Technical review and analysis reports on selected proprietary documents
- Preparation of the internal standard design manuals and procedures
- Establishing the design Quality Assurance system to acquire the ASME N-stamp certificate
- Install, verify and maintain all design computer codes imported from CE to be fully operational at all times
- Operate the Design Document Control Center (DDCC) at Daedeok to sort, store, and distribute.

This extra effort was the reserve manpower trained for the coming new projects like Ulchin 3&4 and Yonggwang 5&6. In fact, what may appear to be excessive manpower loading at the beginning, turned out to be the exemplary program of manpower training and build-up to start from CRT to OJT to JSD for the very first project YGN, then moving on to subsequent projects. This was all possible due to the fact that Korea kept on building the standardized fleet of twelve more NPPs during the last twenty years: Yonggwang 3&4, Ulchin 3&4, Yonggwang 5&6, Ulchin 5&6, Shin-Kori 1&2, Shin-Wolsong 1&2, all of the same design Generation II NPPs (the suffix "Shin" means "new" in Korean, thus "Shin-Kori" refers to the new NPP site adjacent to the "old Kori" site). The ultimate sign of technical self-reliance comes from the ability to take up the warranty responsibilities for the performance and delivery of the new NPP. While CE was responsible for the warranties for the YGN, KAERI (later KOPEC) was responsible for all subsequent system design projects from Ulchin 3&4.

Mission "95-by-95"

On December 7, 1995, a quiet celebration was going on at the KEPCO head office in Seoul following the technical self-reliance evaluation symposium. It was declared by internal as well as external experts that the dual project milestones of Yonggwang Unit 3 going into commercial operation and the technical self-reliance goal of 95 percent had indeed been reached. The component design and equipment supply of the reactor coolant pump, man-machine interface system in the main control room, and the third-party-owned computer codes were the only remaining 5 percent yet to be localized. The unique strategy of the joint design seemed to have worked to overcome the shortages of experienced manpower, budget, and time. We all experienced a big sigh of relief that the major milestone was achieved. At least we knew we could replicate the design of the same NPPs indigenously. The next challenge was how to move up to a truly Korean reactor model design to compete with the Generation III models being developed by other international suppliers. Every indication was positive in that we were ready to take up the next challenge: the Korean Next Generation Reactor (KNGR) development.

From the day one of the very beginning of the YGN project, KEPCO announced the goal of reaching 95 percent of technical self-reliance for the entire NPP plant design and construction by the year 1995, when the Unit 3 was to be completed, thus the nickname "95-by-95." It wasn't clear at all at the beginning what was meant by "95 percent self-reliance" and how to measure its progress. Thanks to KEPCO's rigorous system of accountability in project

The Joint Design

management, the project management team from EPGCC was able to agree on the formula with relative weighing factors that could result in an evaluation of the 95 percent goal by the year 1995. The following table was the agreed formula:

Category	entity	weighing factor	self-reliance target(%)
Total project mg't	KEPCO	15	98
Plant design	KOPEC	21	95
NSSS system design	KAERI	7	95
Initial core design	KAERI	2	100
NSSS equipment	KHIC	24	87
T/G equipment	KHIC	11	98
Fuel fabrication	KNFC	3	100
Construction	Hyundai Const.	17	100
Total		100	95

It is interesting to note that the fuel design/manufacturing, and site construction technologies were deemed 100 percent self-reliant by 1995 due to their past experiences in the existing NPP construction and reload fuel projects. Relative weighing factors were agreed upon primarily based on the contract prices covering the entire spectrum of nuclear power technologies in design, manufacturing, and construction. Each nuclear entity (which formed the EPGCC from 1985 on) knew for sure that they were in for the most challenging episode in the company's history, as each one was assured of exclusive rights and responsibilities in the standardized fleet of NPPs to come.

One of my prime responsibilities as the project manager of system design was to set the annual detail implementation plans for technical self-reliance and confirm it for the monthly progress checks. Numerous internal and external meetings were held at Daedeok, Changwon, and Seoul to discuss, modify, and argue with other entities. It seemed like a nightmare scenario in those days to trace thousand of different design activities and the status of where we lacked confidence. In hindsight, it was the necessary step in gaining confidence, and, most of all, beginning to comprehend the scope of our shortcomings, that is, what we really did not know. There was the natural process of gaining the "know-why" beyond the "know-how" by repeating the nearly identical NPP construction projects and having to defend the adequacy of the design to KINS and KEPCO for each new project to reconfirm the application of the previous project design and to justify the changes that were incorporated into the new projects.

In parallel, the Technology Transfer contract was implemented to 1) transfer documents, codes, patent rights, 2) conduct a series of training courses for the system designers, 3) separate training for the supervisory capacity and project management, 4) procure special computer control equipment. Additional efforts were made on the System 80+ R&D projects together with CE to enhance further understanding of the know-whys leading to the next-generation power reactors to come. In that sense, the mission "95-by-95" was accomplished over and beyond the original expectations. There was a good reason to celebrate on that night of December 7, 1995.[5]

CHAPTER 12: GROWING PAINS

Safety and regulatory licensing always takes the center stage for most nuclear projects anywhere. The case of YGN was more dramatic than previous Korean projects since it broke many records in many "firsts" in Korea. It was the first NPP project in which the local entities took the prime contractor's roles. It was the first (and the last) project where the local entities evaluated and selected the technology partners from the US. It was the first (and the last) project to have parallel technology transfer contracts. It was the first project of what was to become the fleet of Korean Standardized Nuclear Plants. It was the first project during which the fully independent nuclear safety entity, the Korea Institute of Nuclear Safety (KINS) acting on behalf of the nuclear regulatory body, the Ministry of Science & Technology, came into being. And, most of all, it was the first project built after the 1988 Seoul Olympics, which was a historic turning point in Korea for democracy. Regulatory licensing-related milestones can be summarized as below:

Dec. 1981	Nuclear Safety Center (NSC) established in KAERI
Apr. 1986	Chernobyl accident
Dec. 1986	First KAERI system design team sent to Windsor
Apr. 1987	YGN contracts signed
Mar. 1988	NSC staff sent to Windsor for ten-week system design training
Oct. 1988	National Assembly interrogation on safety issues (Thermal/hydraulic relative size issue)
Dec. 1989	Construction Permit granted with conditions

Feb. 1990	NSC transformed into KINS, fully independent from KAERI
Sep. 1994	Operation License granted with conditions (Safety Depressurization System issue)
Mar. 1995	Unit 3 commercial operation
Dec. 1995	95 percent technical self-reliance declared
Dec. 2010	KINS celebrated its twentieth anniversary

Independent safety licensing

By the end of the 1980s, everyone in the project was keenly aware of technical self-reliance in the regulatory licensing arena being just as important as the NPP design and construction itself. The YGN project has special significance to the national regulatory authority as it was setting records to be the first in many ways, while the project period overlapped one of the most transforming eras in the Korean democratization period following the Seoul Olympic games. The origin of the nuclear safety in Korea can be traced back to 1980, when the concept of the Nuclear Safety Center was formed from KAERI's closure episode (described in detail in Part I, Chapter 2). As KAERI got more deeply involved in commercial nuclear power projects in the mid-1980s, inherent conflicts of interest were growing inside KAERI between the NSC and the commercial projects, including nuclear fuel localization and the reactor system design activities. One side of the house was promoting nuclear power through fuel localization projects and the YGN system design project, while the other side in the NSC became increasingly frustrated as the regulatory side supporting the Ministry of Science & Technology.

After much internal discussion and deliberation within KAERI, together with political pressure building up to strengthen the NPP safety regulatory process, the NSC branch of KAERI finally became independent from KAERI as a separate legal entity reporting directly to the Minister of Science & Technology in February 1990. The new entity was named "Korea Institute of Nuclear Safety (KINS)", to demonstrate a more independent regulatory regime in Korea. Lee Sang-hoon, the director of NSC, was named the first president of KINS, together with some eighty technical staff members relocated from KAERI to be the founding members of the new regulatory body. The Construction Permit (CP) for the YGN project, a major NPP construction milestone to pour the first concrete for the reactor building base mat, was issued in December 21, 1989, fully expecting the new regulatory regime would become independent shortly. The YGN project provided invaluable opportunities for the technical self-reliance of the safety regulatory capabilities in Korea, now further enhanced by the independent KINS. Over eleven hundred technical safety-related questions and answers were seriously treated during the course of reviewing the Construction Permit. It was the first time the indigenous regulatory body exercised its full authority and responsibilities from the beginning of a new NPP construction. It was the dawn of a new era in nuclear safety in Korea.

Koh Byung-joon, then the director of the Safety Review Division and the project manager for the YGN licensing, and Lee Seung-hyuk, then the project engineer, were the most responsible officers from the KINS. I asked Koh what he remembers the most when I met him recently at the KINS twentieth anniversary ceremony in Daejeon:

> "To be sure, Yonggwang was the first opportunity for KINS to prove the technical competency of an independent national regulatory body. I still remember vividly the technical arguments after arguments on some major issues. We had to satisfy ourselves to get to the bottom of the issue. I feel proud of having been able to manage the challenge, together with my colleagues at KINS."

When the Construction Permit for the YGN was issued in December 1989, it was given with two major conditions: one on the safety assessment of the scaled-down hybrid reactor, and the other on the rapid depressurization capability under accident condition. Both licensing issues left significant impact on the newly formed regulator KINS, as well as the licensees in implementing the design and construction, as described below.

"Thermal-hydraulic relative size"

When the first National Assembly's audit session was held in October 1988, only a few weeks after the Olympic games' closing ceremony, the number-one target was to probe the safety of the YGN project, led by well-known opposition party leaders like Whang Byung-tae. Whang and others made allegations that the safety of the NPP could not be assured since the basic system design was a "scaled-down hybrid" version and thus not proven, and further implying political slush fund irregularities. Suddenly the most ambitious and proud project was on the verge of plummeting into an unsafe, unethical political scandal. Whang demanded we should obtain the USNRC license certificate to prove its safety. His argument was that the YGN system design was only a scaled-down version of the proven Palo Verde NPP in Arizona (which was 1300 MWe size), combined with the core from ANO-2 NPP in Arkansas (with 1000 MWe size, the same as YGN). The so-called "scaled-down hybrid reactor design" of the YGN NPP was severely criticized in the Korean media for not having been proven. All safety-related probing questions from the audit were in the area of the reactor system design, putting me and my KAERI colleagues on the spot.

An official inquiry letter was sent from Korean national assemblyman Kim Jong-shik in August 1988 to Lando Zech, then the chairman of the USNRC, stating:

"... We are concerned as to whether our present CE design for Yonggwang 3/4 (the combination of scaled-down System 80 and pre-System 80) would be licensable or not in the US..."

USNRC Chairman Zech's response letter came back with a polite refusal, as expected, stating:

"As you are probably aware, NRC does not conduct evaluations of the safety of nuclear power plants exported from the US to another country. Such evaluations have been deemed to be the responsibility of the recipient country..."

However, USNRC gave the names of four US nuclear national laboratories (Argonne, Brookhaven, Los Alamos, and Idaho National

Engineering Laboratory) that could provide technical evaluation of the thermal/hydraulic safety of a hybrid core such as the YGN design.

We made a top-priority technical review contract through NSC and CE with the Idaho National Engineering Lab in the US to make a safety assessment of the Yonggwang reactor system composed of the scaled-down version of the System 80 design. This was later known to be the "thermal/hydraulic relative size" issue, which contributed much to the understanding of our reactor system from the inside out for the regulators as well as the licensee. INEL issued a topical report titled "Thermal-Hydraulic Relative Size Effects" in August 1989, with the conclusion that "it was found that the differences between 3817 and 2825 MWt units led to increased margins, …" confirming the basic design assumptions of additional safety margins built into the YGN reactor design compared to its reference plant, Parlo Verde NPP. This report was considered authoritative and unbiased enough to be able to convince the opposition party lawmakers in Korea. More important was the opportunity given to the new system designers at KAERI to understand the basic mechanism of a power reactor in thermal/hydraulic behavior, which is most essential to the safety of a power reactor. In essence, the report brought to light the autopsy of a reactor as no one had done before.[1]

Although the Construction Permit was granted on time, one of the conditions attached to the scaled-down hybrid core issue was to demonstrate that no excessive vibration and wear would be induced. The scaled-down core of YGN was considered a non-prototype category, which required physical vibration testing. USNRC just revised a new Regulatory Guide on such requirements, known as the Comprehensive Vibration Assessment Program (CVAP). After a number of meetings between Windsor and Daedeok for the design and analysis, and Changwon and Newington for the manufacturing of reactor components, additional measurement and vibration testing were conducted at the fabrication shop in the US. Additionally, in-situ measurements at Yonggwang Unit 3 reactor internal were made during its hot functional testing. To everyone's relief, the final inspection of the internal after the test showed no sign of excessive wear or tear. It took almost five years to address the "scaled-down hybrid" issue to everyone's satisfaction, from a US national lab's analysis to the ultimate in-situ vibration testing at the Yonggwang site. It was the first major licensing lesson learned from a newly independent regulatory body in Korea.

"Chang's valve"

Perhaps the single most important licensing issue throughout the entire project period could have been the safety depressurization system (SDS) case, which led to the permanent physical hardware changes on the pressurizer and the decay heat removal system. One may recall one of the TMI follow-up action items was to eliminate (or replace) the pressurizer relief valve. It provided the direct cause of the TMI accident in 1979 when the valve was stuck open and not closing. An immediate cure was to eliminate such a valve at the top of the pressurizer from the design stage and NPPs built in the early 1980s, such as the Parlo Verde plants, which had no pressurizer relief valves. Parlo Verde being the reference design for the YGN, it was natural to expect the same design as the project began in 1987.

The world nuclear community was developing a new requirement to address a more fundamental approach to severe accident analysis and design of NSSS in the 1980s. Since there were not many new NPP construction projects on-going in the 1980s and 1990s following the Chernobyl accident, YGN was the only new project under construction with the US design origin. A new concept of safety depressurization system together with a feed-and-bleed system at the pressurizer was emerging as a new industry standard adopted to the ALWR design in the US (and later adopted in the System 80+ design of CE). IAEA's nuclear safety advisory board to its Director General, International Nuclear Safety Advisory Group (INSAG), had its collection of international safety experts' opinions aimed at the adoption of SDS in all new NPPs. If this new SDS were to be adopted at the YGN project, it would have created a major back-fitting of both units' pressurizers,

Growing pains

which were already installed at the site. This meant additional significant project cost increase and possibly a schedule delay. Pressure was rising at the project review meetings and licensing meetings among KINS, KEPCO, KAERI, CE, and outside experts groups. It was a small battle in 1991 between the newly born Korean licensing authority at KINS who pushed for the installation of the new SDS, while the project teams around KEPCO were naturally reluctant to accept yet another major design change.

Chang Soon-heung, a professor of nuclear engineering at KAIST in Daejeon, could be the one person most responsible for the resolution of the SDS issue with Yonggwang licensing. Chang, PhD graduate of nuclear engineering from MIT, specializing in the severe accident thermal hydraulics, returned to Korea in 1982. This was the time when the new technical self-reliance drive was just initiated with the onset of the YGN project. He began to show his involvement on the new NPP project as a member of the Advisory Committee on Reactor Safety (ACRS) to the KINS licensing activities in his capacity as a severe accident specialist. In addition, he served as a member of the INSAG at the IAEA during the 1991-1998 period, when the Yonggwang SDS issue was at its peak. I asked Chang about his conviction almost twenty years ago and how he could be so sure to insist on such a design change, coming from a college professor's position:

> "YGN was the first real NPP project under construction since the TMI accident to address the severe accident issue. I was fully aware through the INSAG members that a new rapid depressurization mechanism like the SDS would be a mandatory requirement for all future ALWRs. The USNRC was about to issue a new Regulatory Guide on it. Since the YGN was the onset of many more KSNPs to follow, if we did not succeed at Yonggwang, we would never be able to enforce it later."[2]

Time and again Chang made stubborn appeals that the YGN units must have the depressurization valves to address the residual decay heat removal under a severe accident scenario. He even persuaded the CE system design managers, including Regis Matzie and Tom Natan. It was known as the "Chang's valve," and the question was "to have" or "not to have" the Chang's valve for the Yonggwang project. KEPCO, together with KAERI, CE, and KOPEC, was in a bind in 1992 over what to do next. It would take a considerable amount of back-fitting to the already installed pressurizer; however, the risk was also high in the Operations License stage to delay the commercial operation schedule as long as the Chang's valve issue was not fully resolved. At this crucial juncture, the KEPCO's top management, President Rieh Chong-hun and the YGN Project Manager Park Yong-taek, made a strategic decision to go ahead with the SDS design change, but strictly as a voluntary measure rather than a formal licensing follow-up action. It ended up saving face for both sides; the regulatory body was assured that Chang's valve was installed, while KEPCO won confidence and applause from inside and out to be a forward-looking, progressive electric utility company. It was taken as a win-win situation for everyone involved. Thus, the YGN units have installed the SDS system as a voluntary in-situ back-fitting, and distinguished themselves from the Parlo Verde plant, which was the reference plant. It is comforting to look back on what happened nearly twenty years ago: the SDS system reduces one order of magnitude of severe accident frequency, judging from the probabilistic safety analysis. The YGN units were the first NPP worldwide to have the SDS system built-into the pressirizer; it is the most distinct feature of KSNP design compared with the reference System 80 plant, as built in Palo Verde NPP, and the APWR designs today have the rapid depressurization system such as the SDS as a standard design.[3]

Integrated Safeguards

The time of the YGN project also coincided with the most volatile period for nuclear diplomacy in the Korean Peninsula, which was totally out of control from the perspective of the project team. Immediately following both North and South Koreas' formal admission to the United Nations membership on September 17, 1991, President Roh Tae-woo's government made a bold political move to reach an agreement with North

Korea on the non-military use of nuclear energy on the last day in 1991. This unilateral statement, known as the Joint Declaration of Denuclearization in the Korean Peninsula, was signed by Prime Ministers Chung Won-shik and Yon Hyung-muk of South and North Korea, respectively. Main highlights of the Joint Declaration were: 1) both Koreas shall not develop any nuclear weapons, and use nuclear energy solely for peaceful purposes, 2) both Koreas shall not possess reprocessing and enrichment facilities, and 3) both Koreas establish the Nuclear Control Joint Committee (NCJC) to implement mutual inspection of nuclear facilities. The good intention was to promote peace and security in the Peninsula by prohibiting North Korea to move toward its nuclear weapons ambition, as it coincided with the timing of the North Korean ratification of the NPT and a full-scope safeguards agreement with the IAEA. History speaks for itself that this declaration was unilaterally abandoned by the North in the subsequent years. Interestingly enough, however, it left a lasting legacy to South Korea to better equip itself in the nuclear nonproliferation arena.

As stipulated in the Joint Declaration, both Korean governments initiated the implementation of the mutual inspection by organizing the NCJC immediately in 1992 without knowing what the nuclear inspection, meaning the nuclear material accountancy control, was all about. A series of NCJC meetings were held at Panmunjum (a famous border-town in the Demilitarized Zone, mere forty kilometers north of Seoul) to discuss detailed procedures of getting the mutual inspection started. Subjects of the nuclear sites, inspector selection, and how to inspect and report were discussed, but all ended in vain as the North was demanding military sites to be included. In hindsight, it seems apparent that the North never had any real intention to open their nuclear facilities for inspection from the South; it was only a political gesture. Nevertheless, a multi-ministerial task force team was organized from MOST, the Ministry of Defense, Ministry of Unification, and KAERI to prepare and train the nuclear inspectors to be sent to the North should the mutual inspection take place. I was among the dozen nuclear experts drafted to join the task force team as a reactor specialist. We knew there are several graphite reactors, fuel fabrication, and reprocessing plants at Yongbyon Nuclear Center in the North, which was the primary target to be inspected. South Korea had virtually no technical expertise in the area of safeguards or nuclear material accountancy control in the early 1990s, as it was considered a nuisance and misunderstood issue of nonproliferation, only for the IAEA to conduct the safeguards inspection. This mutual inspection episode helped to change all that perception. The task force team was put through crash training courses with the help of the US government. Special class-room lectures were organized at the KAERI site with hands-on experiments using many inspection tools and neutron detectors. The task force team was sent to the US national nuclear labs at Oak Ridge and Los Alamos National Labs for simulation training of graphite reactors and reprocessing lines. By mid-1993, the task force team was fully trained and mobilized, and ready for the D-day of the mutual inspection mission. It was a bizarre experience for a nuclear technical person like myself to be ready (mentally as well as physically) for a possible mission to Yongbyon, North Korea, of all places on earth, which was a totally isolated secret place to the outside world. Fortunately (?) this mission never materialized as the North abandoned all known forms of international and bilateral nuclear treaties and agreements in the following years of the North Korean nuclear crises. (This subject is dealt with in detail in Part I, Chapter 7, "Lessons learned the hard way.")

History has a strange way of making headway, and even changing one's personal career in a most unexpected manner. Even though this episode of the mutual inspection in the early 1990s turned out to be an impossible dream, it gave an eye-opening experience to the world of nuclear nonproliferation, which was considered "taboo" to many Korean nuclear entities at that time. If we were to move into a higher level of nuclear technologies, including sensitive fuel cycle technologies, it made sense to have our own nuclear material accountancy control technology, but was a far neglected area in Korea. A bit more ambitiously, if we were to get into a more serious nuclear trade with other countries, such as nuclear plant exports, it seemed desirable to have this genre of technology to be established after all. Accordingly, President Shin Jae-in made a decision to establish a new Technology Center for Nuclear Control (TCNC) at KAERI in 1994. Its mission was to set up a national infrastructure of State System of Accountancy Control (SSAC) for South Korea in conjunction with the IAEA inspection scheme under the full-scope safeguards agreement. I was given the opportunity to serve as a director of TCNC (1995-2000) while acting as a member of SAGSI (Standing Advisory Group in Safeguards Implementation, advisory role to the Director General) at the IAEA. In one aspect, it opened a door to the new area of nuclear technology in the nonproliferation and disarmament disciplines, which were by and large neglected in South Korea until the mutual inspection with North Korea came up.

The technology in the nuclear material monitoring and accountancy control has its origin in the early 1970s in the US when the Nuclear Nonproliferation Treaty (NPT) regime was created, followed by the full-scope safeguards agreements being implemented with the IAEA in Vienna. Key scientists from the US weapons laboratories, such as Los Alamos and Sandia, founded a new technical society called the Institute of Nuclear Materials Management (INMM) in 1972 to find links with the civilian use of nuclear technology in dealing with nuclear materials such as uranium and plutonium. Today INMM provides a forum of international exchange of technical information in the nonproliferation arena with several international chapters. South Korea created the INMM-Korea Chapter in 1997 after the TCNC was founded to join the global technical community to enhance its capabilities.

When South Korea was planning the mutual inspection with the North in the early 1990s, the IAEA was deliberating the new strengthened safeguards regime, known as "Program 93+2" (negotiations were started in 1993, and were hoped to conclude in two years), following the revelation of undeclared clandestine nuclear activities in Iraq and North Korea. It took much longer to reach international consensus to give the IAEA broader mandate to inspect undeclared nuclear activities as well as the conventional declared ones. This is known as the "Additional

Protocol" to the existing safeguards agreements with the IAEA, today the *de facto* international gold standard of safeguards inspections. Once the national nuclear inspection program was established at the Ministry of Science & Technology with TCNC support, South Korea signed the Additional Protocol in 1999 and ratified it in 2004 in anticipation of the enhanced nuclear R&D and domestic nuclear power program, leading toward an export market.

Since the TCNC was initially created as a division of KAERI for its technical resources at that time, inherent conflict of interests was bound to come up. KAERI proper is the lead nuclear R&D center dealing with nuclear materials for research while the TCNC's mission was turning into a regulatory function to monitor and control nuclear materials in conjunction with the IAEA safeguards inspection scheme. A critical moment came in 2004 when the voluntary initial reporting on the nuclear materials history at KAERI was submitted to the IAEA for its Additional Protocol ratification follow-ups. It contained several unreported laboratory experiments dealing with plutonium and uranium separations, which happened many years ago. The quantities of nuclear materials involved were minute (in fraction of gram orders) and the intention appeared to be purely scientific in nature. However, the news media printed headlines on the suspected clandestine nuclear activities. The South Korean government was most embarrassed to admit that it was a clear mistake not to report to the IAEA on time. Some even concerned that the case could be brought up to the UN Security Council. A special agenda on the "Implementation of the NPT safeguards agreement in the Republic of Korea" was discussed at the IAEA Board meeting on November 25, 2004. Ambassador Choi Young-jin (then the Vice Minister of Foreign Affairs) came to the Board meeting to deliver a moving speech on this embarrassing agenda for his country:

> "Some twenty NPPs are in operation in Korea delivering over 40 percent of its electricity demand. Such a country had everything to lose and nothing to gain by disturbing the inter-

national peace and security regime, especially the nuclear non-proliferation regime.

The important question was one of intent, i.e. whether or not the government had been involved. It had not been and there the facts spoke for themselves. The Republic of Korea had signed its Additional Protocol in 1999 and the AVLIS (Atomic Vapor Laser Isotope Separation) experiments involving uranium had taken place in 2000. Would any sensible government have allowed undeclared experiments to take place one year after signing the Additional Protocol, knowing that such activities shortly be subjected to the Agency inspections? Of course, lack of intent did not absolve the Government of its obligations. However, the reporting failure in question had taken place in the context of isolated laboratory-scale experiments, not as part of any deliberate program to enrich or reprocess nuclear material, let alone to turn it into weapons. My country's case was therefore technical in nature, not military or security-related."[4]

This was indeed a moving and convincing speech to the international forum from the chief diplomat of a country under suspicion. I was sitting in the Board room in that moment as an IAEA director, and personally felt proud to have such a diplomat speaking up for South Korea on this critical issue. Consequently, the South Korean file, known as the "nuclear materials hiccup case," came to an end at this Board meeting. However, a much bigger change was waiting at the home front. The much-debated issue of creating an independent nuclear non-proliferation agency separate from KAERI was finally put into action. New legislation was passed to create an independent entity, Korea Institute of Nonproliferation and Control (KINAC), in 2006, reporting directly to the Ministry of Education, Science & Technology (MEST). A significant milestone was reached at KINAC in 2008 to have the Integrated Safeguards regime formally recognized by the IAEA. It enables efficient implementation of safeguards inspections in South Korea in recognition of the advanced status of the national inspection system. The IAEA inspection efforts could be reduced to half (about sixty person-days-inspection from over one hundred) while the KINAC's national inspection was being increased. A lesson was learned the hard way after much national embarrassment, but left a lasting imprint in the South Korean nuclear community of the importance of nuclear transparency to the world. An era of nuclear exports, further developments in the nuclear fuel cycle research, and the ultimate solution to nuclear-armed North Korea would require further a significant role of KINAC in the coming years. In addition to the national safeguards activities, KINAC's mission covers the export control and physical protection in line with the expanding nuclear program. Nuclear security, particularly in terms of regional manpower education and training, is being enhanced in preparation for the upcoming Nuclear Security Summit in Seoul in 2012.

CHAPTER 13:
NEXT GENERATION REACTORS

The preceding Chapters in Part II of this book, "Know-Hows and know-Whys," detail the process of how the initial technical self-reliance of nuclear power technology was reached during the first three decades. The pinnacle of this effort came in the mid-1980s with the grand technology transfer in the YGN construction project. The so called "95-by-95" goal was achieved with the ability to virtually copy the same KSNP design reactors over and over again by the time Yonggwang unit 3 went into commercial operation. They may not have understood all the "know-whys" but managed to learn and copy the "know-hows." It was more than anyone could hope to achieve at that time, but that was not to be the end. An additional ten units of KSNP reactors were built back to back in the next decade: Ulchin 3&4, Yonggwang 5&6, Ulchin 5&6, Shin-Kori 1&2, and Shin-Wolsong 1&2. The Yonggwang site on the west coast was now full with six PWR units in operation, while the three east coast sites - Kori, Wolsong, and Ulchin - were further expanded to the adjacent new sites to be called Shin-Kori, Shin-Wolsong, and Shin-Ulchin sites for future expansion. When you repeat the same design projects under the standardized fleet of NPPs, you are bound to master the subtleties of technologies in gaining efficiency and economics, and most of all, the confidence of the "we can do" spirit. You ask the right questions - "Why does the design have to be done this way?" - moving toward the know-why portion. It took an entirely new approach to an NPP design, starting from the conceptual design to the project implementation. Time has come for the Korea Next-Generation Reactor (KNGR) development program.

Even the thought of exporting Korean nuclear plants to some emerging country was not an impossible dream. This was the reason behind dropping the familiar term "KSNP" to be replaced by the "Optimized Power Reactor," or "OPR1000," being built and operated in South Korea. The message was clear: "We are ready to export the OPR1000" when the right time comes. They were determined to prove the "optimized" KSNPs built in the 1990s would be competitive in the world market with the proven records of safety and economics at home. (see "Korean Nuclear Power Plants Chronology" in page iv.)

When it comes to exporting an NPP that was Koreanized from its US origin to a third country, commercial terms and conditions on the royalty payment becomes a critical issue. The original Technology Transfer contracts with CE in 1987, and the subsequent Licensing Agreement in 1997, all had continuing royalty clauses on the imported technologies diminishing to zero when the Korean reactor was to be exported to a third country (except the USA) after ten years. Although the commercial contract terms were quite favorable to export cases, Koreans needed to do some soul-searching: what do we mean by "Koreanization," how much technology was of US origin, and how much technology was added from Korean origin? From what level of development can we say it is truly a Korean reactor? That was the question.[1,2]

Why KNGR now?

If the initial phase of Technology Transfer from CE in the 1980s was to master the know-how of how to replicate the same design over and over again correctly, the need for understanding the basic know-why's was growing internally to be able to create a truly Korean model reactor. The Ministry of Science & Technology (MOST) initiated the "Highly Advanced National Projects" (otherwise known as 'G7' projects) program in the early 1990s in preparation for the coming twenty-first century. Government officials were particularly keen on future technology trends and which ones could lead Korea into the next century. 'G7' was the symbol of the most advanced economic power countries in the world (US, UK,

Germany, France, Canada, Italy, and Japan), and MOST was leading interministerial task forces to form long term technical projects in high-tech areas, including electronics, telecommunication, biotechnology, and energy. In-depth feasibility studies on technical, commercial, and financing aspects were carried out on each candidate project. High-definition TV and broadband Internet were among the selected top twenty candidate G7 projects.

The nuclear power sector was in desperate need of a basic R&D program as the country was committed to a large fleet of NPPs based on US technologies. The next-generation power reactor development program was formulated at the right time in history to be one of the G7 projects. After all, the existing NPPs in operation and construction were all Generation II reactors developed in the 1970s and 1980s, before the TMI accident. No new NPPs were being built in the US since the 1980s. However, the global nuclear industry was undergoing a fundamental change from vendor-driven business to utility-driven model, led by the Electric Power Research Institute (EPRI) in the US. They formulated the Generation III light water reactor requirements, known as the Advanced Light Water Reactor (ALWR) utility requirements document from a total plant approach. It also incorporated the severe accident mitigation scheme to reflect the post-TMI action plans. The Korean Next-Generation Reactor (KNGR) project was officially initiated in 1992 as a G7 project in three phases with the following missions:

Phase I: Top-tier requirement development
(1992-1994)   Reactor type selection
                 Conceptual design

Phase II: Utility requirement development
(1995-1998)   Basic design
                 Standard Safety Analysis Report preparation

Phase III: Design optimization
(1999-2001)   Standard Design Approval from the licensing authority

The same entities from the Yonggwang project were assigned to take up their respective areas:

KEPCO:	Project management and total integration
KOPEC:	Plant design (later to include NSSS system design)
KAERI:	Initially NSSS system design, later to include basic R&D and verification testing
KHIC:	Manufacturability review
KNFC:	Fuel design
CE:	NSSS consultant

In addition, the technical arm of the nuclear regulatory body, Korea Institute of Nuclear Safety (KINS), was brought in as the licensing review function. Each entity designated a separate KNGR project team independent from the ongoing NPP construction projects. Most fortunate was the fact that the project funding for the entire ten year period was primarily financed by the KEPCO R&D fund as the success of the project was entirely depended on the stable and continuous funding assurance.

Phase I started with the definition of the top-tier requirements. The reactor type was selected to be the evolutionary PWR based on the proven KSNP design, but scaled up to 4,000 MW thermal (1,400 MW electric) from the existing 2,875 MW thermal design. The plant design life was extended to sixty years from the existing forty years, with the seismic design requirement of 0.3g (Safe Shutdown Earthquake, SSE) compared to 0.25g. These were the basic design requirements of CE's System 80+ design, which was under the USNRC Design Certificate process (CE received the DC in June 1997). Instead of starting from an entirely new PWR concept, they selected the proven concept of the System 80+ as the logical extension of the KSNP, plus some advanced design features and passive design features, for the next Generation III reactors in Korea. This was also welcomed by CE as Korea would provide a proving ground for the System 80+ for further development and eventual deployment in Korea. A severe accident mitigation scheme was incorporated from the very basic design. It provided invaluable experiences for the Korean entities, especially the system designers at KAERI, to have Generation III reactor experience from the conceptual design stage to the basic design, and finally obtaining the Standard Design Approval (SDA, equivalent to the USNRC Design Certificate) from licensing authority MOST/KINS in ten years' time.

Nuclear technology Mecca

During the KNGR project period, two major nuclear industry restructurings occurred in Korea. KAERI's all commercial NPP projects were transferred to respective industries, together with its manpower, in January 1997. KAERI had been in possession of the commercial nuclear fuel and reactor system design projects for exactly twelve years since the government decision in 1985. It had achieved its original goal of technical self-reliance through the YGN project implementation, and the time had come to return to the national nuclear laboratory's role. However, it did not happen without the pain and agony of having to separate the staff, mostly involuntarily, as the situation dictated the working staff should go with the project. Altogether, six hundred and ten KAERI staff members changed their affiliations overnight: NSSS system design team to KOPEC, initial core design team and CANDU fuel fabrication team to KNFC, and the radwaste management team to KEPCO. In order to support the research activities with the absence of NPP-related projects, the Nuclear Research Fund was created in 1996 to supplement the government funding to KAERI and others. The fund amounted to 1.2 *won* per every nuclear kilowatt hours generated by KEPCO at the beginning. This provided essential funding for many nuclear R&D projects, including the KNGR project. By the time of the restructuring in 1997, KAERI had completed (or under construction) six KSNP units at Yonggwang and Ulchin plus three CANDU units at Wolsong, and the KNGR system design project. One other major industry restructuring occurred at the nation's sole electric utility company in 2001. KEPCO was split into five regional generation companies (all fossil plants) and one national company responsible for all hydro and nuclear generation plants, the

Korea Hydro and Nuclear Power (KHNP) company. KHNP is ranked among the top nuclear electric utilities in the world with an operating fleet of twenty NPPs and ten more units under construction as of 2010. The parent company, KEPCO, retained the distribution and sales plus the overseas marketing responsibilities for its financing capacity.

Once the restructuring hurdles were settled down, each entity had its mission better defined in Phase II of the KNGR program. Local university academics also joined the KNGR development work. Nuclear engineering departments at KAIST and SNU participated in the advanced design features and experimental testing work to provide theoretical support. The final selection of the advanced design features beyond the System 80+ was narrowed down to several major design changes. They included the severe accident mitigation system containing the In-reactor Water Storage Tank (IRWST) to absorb the residual heat, Direct Vessel Injection (DVI) to the reactor vessel, and the fluidic device in the safety injection system to increase the passive safety capacity. These are some of the new design features built into the KNGR to enhance the severe accident mitigation capacity to provide credible protection from TMI-like accidents.

One special note is made of the large-scale experiments conducted at KAERI during the course of KNGR development work. Once the commercial projects were transferred to their respective industries in 1997, KAERI's experimental team was called upon to conduct several major verification tests using the test loop facilities. Simulating the nuclear reactor primary pressure and temperature for normal operation and accident condition requires special expertise and major investments to conduct the tests. KAERI's very first such experimental hot test loop was built in 1982 to verify the locally produced CANDU fuels for qualification testing. The thermal-hydraulic lab was expanded to simulate the full-scale severe accident conditions in order to verify the design codes as well as the design parameters. Major test facilities such as ATLAS (Advanced Thermal-hydraulic test Loop for Advanced Simulation) were built to experimentally verify the advanced design features incorporated into the KNGR design. Many safety-related questions raised during the licensing review period had to be resolved through the experimental data from KAERI. The standard safety analysis report was finalized during Phase II. Subsequently, the Standard Design Approval (SDA) was granted in May 2002 with the licensing review by the KINS. It signaled the birth of a new reactor, APR1400 (Advanced Power Reactor of 1,400 MWe), after ten years of painstaking development and verification work by the Korean nuclear industries, academia, and research institutes.

It is important to recognize that the nuclear brain power engaged in the development of the KNGR program was concentrated in Daedeok Science Town (renamed 'Daedeok Innopolis' in 2005), north of Daejeon metropolitan city. Daedeok Innopolis (DDI) was born in 1974 as a cluster of national research centers under President Park Chung-hee to promote science and technology. Today, DDI occupies over seventy square kilometers (17,500 acres) north of Daejeon, comprising the Daedeok Research Center area, Daedeok Techno Valley, and Daedeok Industrial Center area, with over fifty national/private research centers and over one thousand high-tech venture companies. Five universities are located in DDI, including the Korea Advanced Institute of Science & Technology (KAIST) and Chungnam National University. About twenty thousand research staff members are working full-time at DDI today (among them, over 75 percent hold PhD and MS degrees). It is the single largest technical research and venture cluster in Korea, comparable to the Research Triangle of North Carolina, Silicon Valley in California, and Tsukuba Research Center in Japan.

The national nuclear research center, KAERI, was one of the first to move to Daedeok in the mid-1970s, followed by KEPCO's research center, Korea Electric Power Research Institute (KEPRI) in the 1980s. As the nation's nuclear program grew in the subsequent decades, new nuclear entities spun off from KAERI to tackle independent missions. A nuclear fuel company (now under KEPCO-NF) was formed in 1981 to supply locally produced nuclear fuels from Daedeok. Korea Institute of Nuclear Safety (KINS) was created in 1990 to carry out the regulatory mission independent from KAERI, but was also located at DDI. Reactor system design work initially carried out by KAERI was transferred to KOPEC in 1997, now under KEPCO Engineering & Construction (KEPCO-E&C) Reactor Systems division, located within the KAERI campus. The nuclear materials accountancy control for safeguards and nonproliferation function of KAERI became independent in 2004 and founded the Korea Institute of Nonproliferation and Control (KINAC), with new complex located at DDI. The electric utility's KEPRI center was split, with a new nuclear engineering and technology center called KHNP-NETEC, in 2001 when the parent company, KEPCO, went through restructuring. During the course of KNGR development work (1992-2002), design and engineering work was done by KAERI, KEPCO-E&C, KEPRI, KHNP-NETEC, KEPCO-NF, and the licensing review by KINS. The nuclear engineering department at KAIST also provided research support for the licensing efforts. What is remarkable is that the high concentration of nuclear technical expertise in research capacities from all major nuclear entities in Korea is located in the Daedeok Innopolis. KNGR was the first integrated outcome of such nuclear power research project on a national scale, which culminated in the granting of the Standard Design Approval of the APR1400 in 2002. Daedeok Innopolis is uniquely qualified to be one of the largest cluster of nuclear power technologies in the world today, thus a nickname was born: – Nuclear Technology Mecca of Korea. Technical synergy among the nuclear entities located in a close proximity, together with close human relationships in similar areas of expertise, played a significant role in reaching the goal of creating the first Korean power reactor design in a ten-year timeframe.

APR1400 is born

When the USDOE introduced the Generation IV nuclear energy systems concept to be deployable beyond the 2030 timeframe, they made a definition of previous Generations I, II, and III reactors in a loose sense, as below:

Generation I: early prototype and power reactors in the 1950s and 1960s, such as Shippingport, Magnox, Dresden, and Fermi 1. Korea has no Gen I reactors.

Generation II: large commercial power reactors built since the 1970s and still in operation, including PWR, BWR, CANDU, AGR, and VVER types. These NPPs had an original life of thirty to forty years, now being life-extended to fifty or sixty years. All Korean NPPs up to Shin-Kori 3&4, including the KSNP units, may be categorized as Gen II reactors.

Generation III: NPPs developed in the 1990s with significant advances in safety, economics, and sixty-year design life, satisfying the ALWR requirements. They include ABWR by GE, APWR by Mitsubishi, AP1000 by Westinghouse, EPR by Areva, and APR1400 by KEPCO. All new commercial NPPs to be built are expected to be of Gen III-type reactors until the 2030 timeframe, should the revolutionary Gen IV reactors become a proven reality.[3]

It is important to note the inclusion of APR1400 among the top five Gen III reactors commercially viable today in the world market, which was the direct outcome of the nationwide consorted effort by the Korean nuclear entities during the 1990s. The Ministry of Science & Technology, who is the caretaker of the Atomic Energy Act, incorporated a new concept of an NPP standard design approval into law in its revision of December 2000. It provided a legal basis for the KNGR design outcome to be blessed as a Standard Design Approval independent from any real NPP construction project. In January 2001, the official name of the KNGR reactor was christened as "APR1400," meaning the "advanced power reactor of 1,400 MW" electric class. The Standard Safety Analysis Report (SSAR in English language) for the APR1400 was formally docketed with MOST; the formal Standard Design Approval (SDA) was granted in 2002 after two years of licensing review by the KINS. In reality, the KINS review team was in full consultation with the KNGR development team throughout the entire ten year period. Major licensing issues during the review were focused on the design changes from the KSNP plants on the severe accident mitigation system, including the direct vessel injection and adoption of the fluidic device to bring in the passive safety features. Additional hardware verification tests were conducted at KAERI to resolve the outstanding issues. A number of Topical Reports were incorporated to the SSAR docket. May 7, 2002, was a historic day for KHNP to receive the SDA for the first time anywhere. Never before had the nuclear utility acquired an SDA (or Design Certificate) from its national regulatory authority without a construction unit to simplify the future licensing process. KEPCO/KHNP had every reason to build the new APR1400 for their next 'new build' project, as well as having their eyes set on the export market potential from the early 2000s.

The real benefit of securing the SDA became apparent with the onset of the first APR1400 reactors' construction at Shin-Kori Units 3&4 in 2006. The construction permit was issued by MOST/KINS only twelve months after the PSAR submittal, otherwise it would have taken at least twice as long for the first APR1400 unit to be licensed in Korea.

Abu Dhabi connection

The news out of Abu Dhabi on December 27, 2009, gave me the direct motivation to write this book. The South Korean consortium led by KEPCO had just won the largest single nuclear power plant construction project in history to deliver four state-of-the-art Generation III APR1400 reactors, creating the first NPP program in the United Arab Emirates (UAE); it was also the first turnkey project nuclear export from Korea. More surprisingly, it was the outcome of the international competitive bidding among globally well-known major nuclear vendors like Areva and Hitachi-GE. Immediate response from the international nuclear community was a 'surprise choice.' Even for a local lifetime nuclear zealot like me, it came as a complete surprise at first. The February 4, 2010, edition of *The Economist* had a special article on the nuclear industry analysis, saying,

> "It has caused consternation among the six big firms that have dominated the industry for decades: GE and Westinghouse

of America, Areva of France, and Toshiba, Hitachi and Mitsubishi Heavy Industries of Japan. Suddenly the incumbents are confronted by emerging-market 'national champions' with the full backing of their government – an invaluable asset in a high-liability business like nuclear power... The South Korean consortium, which includes the heavy industry arms of Doosan, Hyundai and Samsung, three of the country's biggest conglomerates, and uses some of Westinghouse's technology, has worked together for decades, building and operating most of South Korea's 20 reactors. It offered not just to build the plant, but also to run them..."

After the initial exhilaration settled down, I felt compelled to write a book in English to tell the world what happened in Korea over the past half-century. Perhaps not too much about Korean nuclear technology and industry was known outside of Korea. Perhaps an insider's account on how "some of Westinghouse's technology" became the Korean APR1400 reactor technology could be a good story to tell.

The Korean government promulgated a new decree making December 27, 2010, the first National Day of Nuclear Safety and Promotion - in short, "Atomic Energy Day" - to commemorate the first anniversary of the historic UAE contract signing in Abu Dhabi exactly one year ago. Han Pil-soon, ex-president of KAERI during the 1980s, was specially honored with the first-class Order of Science & Technology Merit, the Changjo Medal, the highest award given to a scientist from the Korean government in recognition of his unique contribution in nuclear power technology self-reliance (as described in Part I of this book). Also Koh Joong-myung and Shim Chang-saeng (ex-KEPCO senior members) were honored at this ceremony with the first-class Order of Industry Merit, the Geumtap Medal, the highest award given to industrialists for their unique contribution in nuclear power developments.

The Korean contract for UAE could be a new business model for NPP construction in a newcomer country oftentimes without mature infrastructure or an electric utility company to build and operate the new NPPs. Traditional turnkey or component approach contracts from a nuclear vendor with an electric utility are now being revisited in building, operating, and owning the nuclear stations. It is a known fact that KEPCO may be one of the few entities in the world today with the NPP life-cycle capabilities with its consortium; from design, manufacturing, construction, operation & maintenance, and radwaste. Only exceptions are the fuel enrichment and reprocessing services which are strategically excluded for non-commercial reasons.

UAE has been preparing to become a newcomer country since the 1980s by joining the IAEA, followed by signing and ratifying agreements and conventions on safety, safeguards, and security. Bilateral agreements were made with the US, Republic of Korea, and France to promote trade and cooperation. The so-called *123 Agreement* with the US was made in 2008 to address the strict principle of peaceful use of atomic energy without acquiring proliferation-sensitive technologies, like reprocessing and enrichment. Since September 2010, UAE was selected to serve on the IAEA Board of Governors to reflect its growing recognition in the global nuclear arena. Two UAE nuclear entities were established in recent years: Emirates Nuclear Energy Corporation (ENEC), responsible for building and operating the NPPs, and Federal Authority for Nuclear Regulation (FANR), ensuring long-term nuclear safety, security, and sustainability in peaceful uses of nuclear power and radiation. FANR and KINS made a bilateral cooperation agreement in May 2010, as both national regulatory agencies will be licensing the same APR1400 reactors in two countries. In November 2010, ENEC received the license approval from FANR for work at the Braka site, the first and preferred NPP site located some three hundred and thirty kilometers west of Abu Dhabi on the Persian Gulf. ENEC submitted the Preliminary Safety Analysis Report docket (prepared by KEPCO) to FANR in December 2010 for the construction permit. Ground preparation work at the Braka site is in full swing at the time of this writing. Given the credibility gained by the Korean constructors over many decades in the Gulf region, the Braka Nuclear Power Plant (BNPP) project team is expected to reach its goals on time, within budget, and most of all, with quality. After all, the same project teams that constitute the same supply chain of companies for the Korean NPP projects are participating in the BNPP project.

There are several additional supply scopes in the BNPP project, which may go beyond the normal NPP construction project. One is for KEPCO to seek a design certification (DC) on the APR1400 design from the USNRC to promote additional assurance without being a contract obligation. Related entities who prepared the BNPP PSAR docket are mobilized to apply and pursue the DC process as though the APR1400 reactor is to be built in the US. Already two pre-application review meetings were held between the USNRC and the Korean counterparts in 2010. Similar DC processes are underway at the USNRC for the French EPR design and Japanese APWR design. One other national commitment for Korea was to help establish a new nuclear engineering department for the graduate program at Khalifa University in Abu Dhabi. The Nuclear & Quantum Engineering Department at KAIST is leading the effort in preparing the curriculum and dispatching professors to Abu Dhabi, while Khalifa University is recruiting qualified Emirate students as well as securing the accreditation of the new department by the UAE Ministry of Education.

United Arab Emirates and South Korea share unique connections in modern history. UAE became an independent nation from a British colony in 1971, while Korea became independent from Japan just after the Second World War. UAE was formed among the seven individual Emirates on the tip of the Persian Gulf: Abu Dhabi, Dubai, Sharjah, Fujeirah, Ajman, Ras al-Kheimah, and Umm al-Qaywayn. Abu Dhabi and Dubai are the best known among the seven as members of the influential Gulf Cooperation Council (GCC) and the Organization of Petroleum Exporting Countries (OPEC). The land area of UAE is about 80 percent of South Korea, but the population is only one-tenth, about five million (including four million expatriates). Both countries were among the poorest in their early statehood years. The first oil crisis of 1973 changed everything for the UAE when the price of oil quadrupled overnight. UAE was most fortunate to have a visionary leader, the first President of UAE, Sheikh Zayed bin Sultan al-Nahyan (in short, "Sheikh Zayed"). His understanding of the development needs of UAE was astonishingly perceptive and has revolutionized the country into a modern desert miracle, similar to President Park Chung-hee of Korea realizing the Han River miracle. Literally both UAE and South Korea came from the poorest nations to one of the most advanced economies in the same period during the last forty years. In other words, both countries share recent memories of coming "from rags to riches" through hard work and visionary leadership. The recent year-end edition of *The Economist's: The World in 2011* quotes the GDP per head for UAE to be $44,450, about twice that of South Korea at $22,050 (but considering the purchasing-power parity, it comes out about the same at $27,690 and $31,400, respectively).[4]

Braka on the Silk Road

The Ashmolean Museum at Oxford University is the world's first university museum, dating back to the seventeenth century. The museum was reopened in 2010 after major renovation work. I noticed a new, interesting wall map in the main entrance hall showing the ancient world map of the *Silk Road* with a big inscription saying, "*Crossing Cultures, Crossing Time.*" It must have been a new thematic approach, based on the simple idea that cultures interact with and influence one another. Such interaction comes in many forms: in adaptation of religion, transfer of technologies, migrations of people. Ashmolean is an outstanding place to appreciate the cultures that share a connected history, which stretches from Europe in the West through the Middle East and Asia to the Far East, from ancient times until the present day. The ancient Silk Road was the main avenue of crossing cultures from the East to West, or vice versa. The *Silk Road* (originally *"Die Seidenstrasse"* in German) was named by German baron Ferdinand von Richthofen in 1877 from the lucrative Chinese silk trade idea when he made several expeditions to China. It was a network of routes on land and sea connecting the major metropolitan cities in ancient times: Rome, Istanbul, Baghdad, Samar Cant, Xian, and Gyeongju in Korea. Also, sea routes were actively navigated from the Mediterranean, Red Sea, Persian Gulf, Indian Ocean, and Strait of Malacca to Japan. Silk was not the only item. Many technologies were transferred through the Silk Road, including paper-making technology and gunpowder.

Suddenly I realized the countries on the Silk Road route - Turkey, UAE, Pakistan, India, China, Vietnam, Korea, and Japan - all have one thing in common. They are among the most active nuclear power countries in the twenty-first century. Nuclear power technology was invented in the West during the last century. Japan was the first Asian country to import the technology from the US in the 1960s and develop its own nuclear power technology. South Korea also transferred technology from the US in the 1980s and built its own fleet of NPPs, then even succeeded in exporting to the UAE. Pretty soon China may follow suit. There is no question in anyone's mind that the nuclear renaissance is coming in a big way in Asia more than any other place on earth. More than 60 percent of the world population (6.8 billion) lives in Asia today. China and India alone have about 40 percent of the world population, but the consumption of electricity is far below the OECD average. They have the thriving economy, which will demand electricity growth, ability to finance large fleets of NPPs, and clear government commitment to achieve economic development with nuclear power. Just to get the feel for the magnitude of the upcoming nuclear renaissance in Asia, and China and India in particular, the following tabulations from the IAEA of NPP status for Asian countries would be informative. The IAEA reactor database is the most accurate source, as they update every month to reflect the global trend. Categories of NPPs are made in four groups, according to the IAEA standard:

In operation:	connected to the grid
Under construction:	first concrete poured
Planned:	approvals, funding, or major contracts in place
Proposed:	specific program or site proposals in place

All 'planned' NPPs listed are expected to be in operation within the next ten years while the 'proposed' NPPs are to be in operation mostly in fifteen years, according to the IAEA database.

Table: Nuclear Power Status in Asia (number of NPPs as of Dec. 2010)

	operation	construction	planned	proposed	total
Bangladesh	0	0	0	2	2
China	13	26	37	120	196
India	19	6	18	40	83
Indonesia	0	0	2	4	6
Iran	0	1	2	1	4
Japan	55	2	12	1	70
Jordan	0	0	1	0	1
Korea, North	0	0	0	1	1
Korea, South	20	6	6	0	32
Malaysia	0	0	0	1	1
Pakistan	2	1	2	2	7
Thailand	0	0	2	5	7
Turkey	0	0	4	4	8
UAE	0	0	4	10	14
Vietnam	0	0	2	12	14
(Taiwan)	6	2	0	1	9
Asia total	115	44	92	204	455
	(26%)	(68%)	(64%)	(61%)	(46%)
World total	447	65	143	332	987

(Source: WNA website, World Nuclear Power Reactors & Uranium Requirements, December 1, 2010)

The table above shows the clear dominance of four countries who are benefiting from the nuclear electricity today: China, India, Japan, and Korea. The total production of nuclear electricity from fifteen Asian countries (plus Taiwan) is only about 20 percent of the world's total kWhs generated by nuclear power in 2010. However, this picture will change drastically as large numbers of NPPs are under construction (a whopping 68 percent of the world's new builds under construction are in Asia), and this trend will continue for the years ahead (64 percent of the world's 'planned' and 61 percent of the world's 'proposed' NPPs are from Asia). Particular attention is needed for China and India for their future growth of nuclear power in parallel with their expected growth of national economies. In 2010 alone, twelve new NPP constructions started around the world (meaning the first concrete was poured). Eight of these were in China (Fuqing 3, Ningde 3, Taishan 2, Changjiang 1&2, Haiyang 2, Fangchenggang 1, and Yangjiang 3), two in India (Kakrapar 3&4) and two in Russia (Leningrad II-2 and Rostov 4), just to illustrate the point.

When you superimpose the known NPP sites (in operation, under construction, and planned categories only, as some 'proposed' sites are not known yet) on the map of Asia together with the ancient Silk Road routes, a striking connection can be made between ancient times and the twenty-first century. A nuclear renaissance is coming in a huge way in Asia. Goods and technologies in building and operating NPPs will be pouring into the NPP countries through the air, sea, and land routes. Super jumbo jets and ocean container ships and barges will replace the old camels and horse-backs of the Silk Road days. One UAE example will illustrate this point clearly. The Braka NPP project (contract signed in December 2009 between ENEC of UAE and KEPCO of South Korea) is mobilizing a grand scale of manpower, plant design, equipment manufacturing, and site construction. The Braka site is located on the Persian Gulf coast. The project manpower will be traveling on jumbo jets from Incheon Airport to Dubai and Abu Dhabi nonstop. Hundreds and thousands of pieces of heavy equipment produced in Korea and elsewhere will be shipped on ocean freight container ships sailing from Busan and other ports, through the South China Sea, Strait of Malacca, Indian Ocean, then finally to the Braka site on the Persian Gulf. A historic day will come around 2014 when the first reactor vessel for the Braka Unit 1 will be mounted on an ocean freight carrier at

Changwon Doosan plant, sailing all the way to the Persian Gulf site. It is almost exactly following the old Silk Road sea route. The world will witness the *"Nuclear Silk Road"* in action. I was pleased to discover that the central theme of this book connecting "nuclear" to the ancient "Silk Road" was first written by Roh Eun-rae (ex-KEPCO senior manager, who served as president of KNFC) in 2005 in the company magazine. Roh introduced the Silk Road in historical perspective and predicted that someday Korean nuclear merchants may travel on the Silk Road... what foresight![5]

Truly Korean

Anticipating the first nuclear plant export project to be realized, the next major national program to follow the KNGR program of the 1990s was crafted by the Korean government in 2006 for the next ten-year program called "Nu-Tech2015." It was aimed at creating an NPP technology base totally independent from foreign intellectual property rights, attached in patents and know-how. One may recall the national drive to achieve '95-by-95' during the YGN project to achieve 95 percent of technical self-reliance by 1995, when the unit 3 was connected to the grid. The remaining 5 percent left out of the self-reliance program was categorized into three groups: reactor core design codes owned by the third party (mainly the US government, thus the ownership could not be transferred to Korea with the Technology Transfer agreements), reactor coolant pumps (RCP), and the NPP control center Man Machine Interface System (MMIS). These selected items were exempted from the self-reliance program not only due to their state-of-the-art technical complexities but also because economic incentives to localize were lacking due to limited market potential. Thus, the above three items were exclusively outsourced from foreign vendors in the US and Germany (or had use rights only and no retransfer rights for the design codes) for all KSNP (OPR) units and APR1400 units being built in Korea. Even for the first UAE export project, the policy was adopted to outsource the RCP and MMIS from the same foreign vendors. This made strategic sense to bring the recognized foreign nuclear major company like Westinghouse on board with the Korean project.

Nu-Tech2015 was aimed to create the enhanced Advanced Power Reactor design (APR+) to the standard design approval by the Korean regulatory by the year 2015, which could be claimed totally free from foreign technology dependence, particularly addressing the above three items. The same nuclear entities as the KNGR program were mobilized: KHNP to integrate and project management, KOPEC for the plant design including the reactor systems design, Doosan for RCP and MMIS development, and KAERI for the codes development and verification testing. The consistency of bringing the same nuclear entities with their exclusive areas of expertise over two decades of time was honored once again in the APR+ development as it was in the KNGR program in the 1990s and KSNP in the 1980s. The only difference this time was to focus on the specific items for design, manufacturing, and verification testings needed for the standard design approval from the regulatory body. Several critical computer codes for reactor core design and severe accident analysis codes are being developed using full-scale experimental test facilities, such as the ATLAS loop at KAERI, for license approval by KINS. Two major design and analysis computer codes (owned by the US government), FLASH and RELAP, had been used extensively for most of the KSNP and APR1400 projects in Korea. Major effort has been underway at KAERI, with academia support, to create up-to-date integrated codes to replace them. Locally developed SPACE and MARS computer codes are being verified for the ultimate licensing approval by the KINS. Critical hardware equipment RCP and MMIS is being developed at Doosan Heavy Industries, KHNP, KOPEC and KAERI with the full support of two ministries, MKE and MEST, for the standard design approval by 2015. Safety Evaluation Report was issued by KINS on the licensibility of the locally developed MMIS in February 2009, followed by the first time supply contract to Shin-Ulchin 1&2 project in July 2009, the second APR1400 project being built in Korea.

Coming of the Korea Nuclear Instrumentation & Control Systems (KNICS)

Nuclear power plants are generally known to be overly conservative when it comes to adopting the state-of-the-art technologies, particularly in the high-tech digital electronics area. Their inherent adherence to the redundancy principle and multiple safety barrier concepts is maintained from the design stage. Additional licensing requirements from the nuclear regulatory involve myriad of verification and validation processes, not only for the design safety, but also in operability and reliability of the system components. This is why the NPP instrumentation and control systems are generally slow in adopting the advanced digital control systems compared to other industries such as civil aviation, chemical plants and fossil power plants.

MOST of Korean government initiated a R&D project in nuclear power I&C system development in 2001, inviting experts from KAERI and Korea Electrotechnology Research Institute (KERI). Kim Kook-hun of KERI was named the project manager of the newly created KNICS project. Kwon Kee-choon and the I&C team of KAERI joined the project to bring in the system design experiences from the YGN project, particularly in the safety system development including the most critical Reactor Protection System, as well as the Nuplex-80 know-hows developed by Combustion Engineering. After several years of development work in hardware and software engineering, Kim and his colleagues of KNICS team managed to win the confidence of KHNP for realistic opportunities to supply the new NPP projects. Major industries such as Doosan and POSCO ICT, joined KNICS development for commercial deployments by integrating the systems and components, capitalizing on the advanced high-tech IT technologies prevailing in Korea. Most difficult task was to acquire the Safety Evaluation Report from KINS after 12 months of verification, validation and equipment qualification efforts following the submittal of the Topical Report. KHNP made a landmark decision to adopt the indigenous MMIS to the Shin-Ulchin 1&2 project once KNICS cleared the safety licensing approval in 2009, then the supply contract was signed with Doosan accordingly.

The IAEA made an independent technical review of KNICS in 2010 for its adherence to the international safety codes and standard. Doosan has completed its new nuclear I&C plant near Suwon with sufficient capacity to supply multiple KNICS units, in addition to its heavy machinery plants in Changwon.[6] In commemoration of the completion of KNICS technology, Doosan made yet another landmark decision to realize technical royalty payment to KAERI in 2010 in order to acquire full grant of license to use the core KNICS technologies. It is known to be in a multi-million dollar lump-sum royalty plus the running royalty, registering as the first case of local nuclear industry's recognition of intellectual property rights of the national nuclear laboratory.[7]

'Use right' or 'ownership'?

Why was it so compelling to overcome these last items on the self-reliance package when the world is becoming more globally interconnected with the market economy? A logical answer could be found in the case of France, which had technology transfers from Westinghouse with PWR technology in the 1950s. Framatome deployed a large number of the standardized NPPs in France, followed by turnkey exports to South Africa, China, and South Korea. When Areva is exporting EPR power plants overseas, they do not appear to have the export approval case with the US, which is their technology's country of origin. French EPR technology is deemed sufficiently mature and independent enough from the original Westinghouse technology today, thus becoming free to compete with Westinghouse in any bidding situations. In addition to supply for the construction of NPPs, they are able to transfer their own EPR technology to the recipient country as a separate technology transfer package, as done in the Chinese EPR program today. The Korean goal of seeking the standard design approval on the APR+ NPP design is a definite step closer to achieving the milestone of "sufficiently matured and independent" technology recognition from the US origin "System 80" and "System 80+" technology. It is not entirely clear at this time, however, as to how the "sufficiently matured and independent" technology will be determined between the US and South Korea. This could very well be the subject for the highest paid international patent lawyers to legally settle the subtleties.

One aspect of the legal technology rights often-times confusing to many is the subtle difference between the "ownership" and the "use right," as specified in the Technology Transfer agreements. One can "own" technology in two ways: by inventing it and protecting it by the patent rights, or purchasing someone else's technology exclusively to "own," meaning to have acquired all rights to sell or use to anyone without prior consent from the original owner. "Use right" normally means one can design, produce, and sell the licensed products anyplace but are not allowed to sell, or sublicense the "technology itself" without prior written consent of the original owner. In the case of Korean Technology Transfer from the US for the PWR technology, it was the "use right" and not the "ownership" of the technology, as written in the "Grant of License" clauses. Within the "use right" package, however, the Korean contract contained additional conditions favorable to the licensee, including the royalty-free (or waiving of the license usage fee) clauses applied anyplace, including a third country export case (with the participation of the licensor). In addition, the transfer of technologies for operation, maintenance, and safety review aspects of the NPP was granted. The Korean export contract to UAE well illustrates this point. Also, the same reciprocity is applied to Korean inventions or improvements of the licensed technology, as in the case of the KNGR (APR1400). Westinghouse (then ABB-CE) would be obliged to pay KEPCO a royalty of 4 percent in case Westinghouse sells the APR1400 anywhere. This is clearly spelled out in the License Agreement document of 1997, signed between the Korean consortium and Combustion Engineering, which still remains valid today.[8]

Small but SMART

So far in this book, our interest was focused on the nuclear power reactor for electricity generation. The clear trend was toward bigger capacity, from 600 MWe in the early years to 1400 MWe APRs being built today for the scale of economy. This trend is well suited for large nuclear countries with corresponding electricity grid capacity. However, many developing countries with small grid sizes may require nuclear units of smaller size, around the 100 - 300 MWe range. In addition, their energy needs include heat for desalination of sea water, like many countries in the Middle East. The global market for cogeneration of electricity and fresh water from nuclear reactors is growing and real, enhanced by the carbon-free green economy initiatives. The global nuclear renaissance of power reactors bound to impact the small and medium reactor (SMR) market as well. The USA, Argentina, and South Korea are among the first on the frontier in this new genre of emerging nuclear technology, as no one has succeeded in the commercial market yet. A quick industry survey shows several nuclear vendors are developing their SMR models, adopting similar reactor technology, and applying for the design certification from the USNRC (or equivalent) at present. The preferred choice of technology by every vendor turned out to be the integral PWR type, with proven LEU fuel from bigger power reactors. "Integral" means generally the entire reactor coolant loop is designed into a reactor vessel containing the core, steam generators, pressurizer, and sometimes the main coolant pump. Passive safety features are designed into the reactor system as the power level is much lower than the NPPs (about one-tenth) for inherent safety and simplicity in operation. The original integral PWR concept was pioneered in Germany with the nuclear ship *Otto Hahn* with nuclear propulsion in the 1960s. Russians had similar nuclear ships with integral reactors but no one made it a land-based commercial SMR yet, as the global nuclear market vanished since the 1980s.

Korean initiative on a SMR development started from 1997, when the commercial NPP projects were separated from KAERI to KOPEC and KNFC under the nuclear industry restructuring scheme. This was a turbulent time to transfer over six hundred KAERI staff members to respective industries. However, KAERI management strategically decided to retain twenty some experienced engineers from the power reactor system design group to be the core members of a new SMR development team at KAERI. Kim Si-hwan (then the project manager of the PWR fuel design project, who later served as vice president of KAERI) was the principal architect of the new development project. The new project was given a name, SMART (System-integrated

Modular Advanced Reactor), exclusively for the export market for the cogeneration of electricity and water from desalination. Clearly there was no domestic need for small-size NPP and fresh water production within Korea, and a prospect of nuclear plant export must have appeared an impossible dream for many in those days. Kim Si-hwan played an important role not only in motivating the SMART team but also to solicit external support from the government. His role as the chairman of the IAEA advisory group INDAG (International Nuclear Desalination Advisory Group) was useful in gaining domestic support and international recognition. During the next eleven years, the SMART project proceeded with conceptual and basic design, analysis and verification work at KAERI, despite several episodes in the project acceptance and funding upheavals with the government. Finally, the government approval to proceed in full force came in 2008, to obtain the Standard Design Approval by the end of 2011. This time the probability of success was heightened with the participation of an industry consortium headed by KEPCO and supported by industry giants like POSCO steel and STX shipbuilder for future deployment. This was the challenge to be met by the Korean nuclear entities: to design and construct a truly one-of-a-kind engineering product for commercial operation from scratch. Home-grown nuclear power technology from KSNP in the 1980s and APR1400 in the 2000s has gone through a logical extension to create something truly Korean: original nuclear power technology in SMART. There is no reference plant to speak of, no one to lean on. Koreans have gained enough confidence for the challenge, from the unproven new reactor design to the final implementation.

SMART system designers at KAERI (about one hundred full-time staff) are in the midst of a three-year standard design verification project (2009 - 2011) to obtain the Standard Design Approval from the regulatory body, KINS. The standard model is of 330 MW thermal output (90 MW electric plus forty thousand tons per day of fresh water), a sufficient size to support the electricity and water needs of a city with a population of one hundred thousand. Design analysis and verification works are ongoing for major equipment and design codes development using integrated thermal/hydraulic test loops at KAERI. A Standard Safety Analysis Report (SSAR) was submitted at the end of 2010 for the licensing review. Eleven Topical Reports have been prepared for addressing unique design features and analysis methods. The know-how's and know-why's accumulated from the NPP self-reliance program of the 1980s - starting from the imported System 80 technology, then to the KSNP deployments, to KNGR development, and to APR1400 approval has finally come to the SMART realization over the past quarter of a century in Korea. It is a risk worthy of taking for a bigger return, but the final outcome remains yet to be seen.[9]

Supply chain

When Korea was importing NPPs from the US, Canada, and France in its early days (1970s -1980s), the vast majority of components and equipment were imported from the countries of origin, particularly for the nuclear-safety-related items. Thousands of mechanical, electrical and civil structural components were designed, fabricated, and certified by the respective national authorities for the nuclear components. International codes and standards, such as the ASME (American Society of Mechanical Engineers), IEEE (Institute of Electrical & Electronics Engineers), ACI (American Concrete Institute) for the Westinghouse-supplied Kori and Yonggwang NPPs, RCC (*Regles de Conception et de Construction*) codes for French-supplied Ulchin NPPs, and CSA (Canadian Standards Association) codes for Wolsong NPP, were adopted. It was natural to apply the industry codes of the countries of origin for the turnkey contract projects; however, local industries experienced the additional burden of becoming qualified bidders as the local contents started to increase in the late 1970s, when KEPCO adopted the component approach starting from Kori Units 3&4.

It took a bold management decision to bring up the otherwise unproven, unqualified local industries to become qualified bidders for the NPP components and equipment. President Kim Young-june of KEPCO (1976 - 1982) was one person responsible for laying a solid foundation in KEPCO to set up the nuclear power program with local industry support. He was a Minister of Agriculture under Park Chung-hee, but left a lasting legacy as a KEPCO president to bring long-term NPP construction plans to fruition. He initiated a system of local industry's specialization scheme to become a qualified bidder for NPP equipment supply chain. As an incentive for local industries to come forward, KEPCO offered guaranteed procurement policy once they qualified. This was a bold move by the national electric utility in the 1970s when the technical levels of the local industries were still in their infancy. In addition, he created a new category of "Nuclear Sector Staff" system to bring higher incentives for those working in the NPP projects. Most talented employees were recruited to the planning, construction and operation of NPPs within the company. No doubt the well motivated nuclear staff at KEPCO contributed to one of the best managed NPPs in the world today.[10]

Korea Electric Power Industry Code (KEPIC) was first created in 1975 by KEPCO with the intention to create the Korean codes and standards for the local industries to be able to meet the stringent safety and quality standards. Lee Chang-kun of KAERI, who later served as an Atomic Energy Commissioner, has been the lead KEPIC chairman from the beginning. Their initial goal was to replace the foreign codes such as ASME and RCC with something Korean to reflect the local conditions. Major heavy industries such as KHIC and Hyundai were already gearing up their plants to receive the coveted ASME N-stamps in the 1970s, as they had strong intentions to move up to the nuclear vendor's status. However, most of the small- and medium-size industries' stories were different. They did not have the infrastructure to be able to acquire higher-level certificates like the ASME N-stamp. A typical valve-maker like Samshin started to supply non-safety-grade valves to the Kori NPP project construction in the 1980s. As an incentive for local industries to take part in the NPP supply chain of some safety-grade components, KEPCO started to offer special sole-source procurement benefits to those who took special licensing arrangements with the foreign supplier at the beginning.

It took nearly twenty years for the KEPIC system of the Korean national industry code to be legally authorized by the Korean government in 1995, in full conjunction and cooperation agreement with the ASME and other international codes. KEPIC codes are updated every five years with annual addenda (similar to the ASME, which updates every three years). They encompass virtually all spectra of industries producing mechanical (about 45 percent of all components), electrical, civil/structural, and nuclear materials and components needed for NPP design and construction. Today well over two hundred Korean local manufacturers and constructors are listed under the KEPIC system, as they are the only qualified bidders for all NPP projects in Korea to form a viable supply chain of nuclear industries. What started nearly thirty-five years ago with KEPCO is now paying handsome dividends in economic and internationally competitive nuclear industries. They are the backbone of the bottom line economics of nuclear power in Korea today.[11]

CHAPTER 14:
RESEARCH REACTOR STORY

Research reactors are used as the principal tools for the development purposes of nuclear science and nuclear engineering as neutron sources for material testing, research, and radioisotope production, while power reactors produce heat to generate electricity. Research reactors are smaller and simpler than power reactors, operating at lower temperatures and pressure. They need far less fuel but require higher enriched uranium, typically up to 20 percent U-235 enrichment fuel, to produce higher neutron flux and power density. There are about two hundred and fifty research reactors in operation today, scattered in fifty some countries, including Africa, a much wider distribution than the power reactors. Research reactors are generally considered as precursors to the introduction of power reactors, as the main nuclear manpower training tool as well as introduction to nuclear sciences. Some major medical/industrial isotopes for cancer diagnosis and nondestructive testing, such as technetium-99 and iridium-192, are exclusively produced from research reactors. All thirty power reactor countries have a large number of research reactors in support of their nuclear power program; however, not all research reactor countries have power reactors introduced yet. The story of the Korean research reactor program, which had its share of ecstasies and letdowns over the last half-century, may shed some light on how it evolved and contributed to its power reactor program. What started out as a simple low-power research reactor imported from the US in 1959, then moved on to indigenously designed high-power HANARO (High-flux Advanced Neutron Application Reactor) thirty-five years later, resulted in the first turnkey export contract of a research reactor to Jordan in 2010. It was the first export project for a research reactor, just about the same time that Korean power reactors were selected for the UAE contract. It was more than a coincidence.

TRIGA legacy

The decision to construct the nation's very first research reactor (KRR-1) was made in 1958 in parallel with the founding of KAERI nuclear research center, in the suburb of Seoul, by Rhee Syngman, the first president of the Republic. Considering that Korea was one of the poorest countries on earth at that time, it was a bold and historic beginning. KRR-1 was a small low-power (only 100 kW TRIGA Mark-II from General Atomic of the US) reactor for simple experiments but its mere presence was enough to motivate two top-ranking universities (Hanyang University and Seoul National University) in engineering to establish nuclear engineering departments for the first time in 1958 and 1959, respectively. A ten-time more powerful research reactor (KRR-2, 2 MW TRIGA Mark-III) was built in 1972 right next to KRR-1 at the KAERI campus to meet the growing demands of nuclear applications in industry, radioisotope production, and manpower training.

During the first two decades, two research reactors at KAERI played an incubating role for training future nuclear leaders in Korea, who later served at responsible positions in all nuclear entities. However, the utilization of research reactors was on the decline from the mid-1970s, as it was difficult to relate the KRR-1/KRR-2 for power reactor simulation for the lack of power density and neutron flux level. Besides, the first series of NPPs built in Korea were all from turnkey-type contracts, requiring little indigenous research and experiments. A real need for a high-power research reactor had to wait until the mid-1980s, coinciding with the NPP self-reliance program, which was activated with the YGN project.[1]

One of the alternative solutions derived from the "close down KAERI" episode of 1980 was to relocate the entire KAERI campus from Seoul to Daejeon (refer to Part I, Chapter 1). When the relocation move was fully activated in 1985, it was found increasingly difficult to operate two research reactors in Seoul. In fact, the Seoul campus lot was sold to KEPCO for their training center site, and KAERI was obliged to decommission both research reactors for other non-nuclear use by KEPCO. The Korean government made a commitment to replace two research reactors in Seoul with a more powerful one in Daedeok when the relocation move was complete. This turned out to be a blessing in disguise: to design and build an indigenous research reactor, HANARO, in the 1990s. KRR-1 and KRR-2 in Seoul were permanently shut down in January and December of 1995, respectively. Decontamination and decommissioning work was commissioned for both reactors, with complete removal of radioactive materials from the core for radwaste treatment and packaging in drums and containers. They are waiting for the final disposal at the radwaste repository site being built at Gyeongju near the Wolsong NPP site. The KRR-2 reactor building was completely decommissioned in 2008, while KRR-1 will retain its original form after decommissioning to be preserved as a historic landmark science museum.[2]

Road to HANARO

Preparation for a high-power material testing reactor with high neutron flux started in the early 1970s at KAERI when Korea was considering the first CANDU NPP from Canada. The road to HANARO took much longer and was more expensive and turbulent than anyone had ever expected. The NRX-type research reactor was considered part of the CANDU Wolsong construction package. The Indian nuclear explosion in May 1974 changed all that scheme since the Canadian government put an embargo on any future NRX exports due to proliferation concerns. Intellectual curiosity coupled with the desire to build Korea's own research reactor indigenously led to the creation of the Thermal Flux Test Facility (TFTF) project at KAERI to continue the basic design from 1976. The principal lead scientist was Kim Dong-hoon (1931-1998, later vice president of KAERI), who was involved in both the KRR-1 turnkey import project and KRR-2 joint construction project with local industry; he led the TFTF project as the expert of research reactor technology. The project continued to the detail design stage with the adoption of natural uranium fuel then came to an abrupt halt in 1981. International nonproliferation concerns were mounting pressure to abandon any research reactor with natural uranium fuel. The Korean government had to cancel the project. However, it left valuable insight in how to approach a new research reactor design indigenously. Several prototype reactor components were developed and tested. It was the first time Korean scientists endeavored in developing a research reactor design from scratch.

Internal soul-searching within KAERI was ongoing in the early 1980s to reformulate the new research reactor powerful enough to be useful in power reactor material testing, as well as meeting the international nonproliferation concerns. The KAERI campus was being relocated from Seoul to Daejeon while the existing KRR-1 and KRR-2 reactors were waiting for permanent shutdown. Having experienced project cancellations twice already, the project team was determined to carry on to the end. A real blessing came in a most unexpected way. Intelligence reports on a new research reactor (estimated to be about 30 MW size) construction at Yongbyon Nuclear Center in North Korea somehow must have convinced the South Korean government to resume the go-ahead decision on the mothballed project. Technical minds at KAERI under Kim Dong-hoon were recalled immediately to start a new project. They knew it would be the only opportunity in their life time to design and to realize a new research reactor once and for all to meet various utilization requirements, including material testing, neutron beam experiments, and isotope production. As such, the project code name was "Korea Multi-purpose Research Reactor" (KMRR) to distinguish it from the previous efforts. An in-depth feasibility study produced a conceptual design with the following characteristics:

- Open pool type (not a tank type)
- Power level of 30 MW thermal output

- Nuclear fuel of 19.8 percent enriched uranium silicide (to meet the nonproliferation requirements, not natural or highly enriched uranium)
- Light water (H_2O) cooled and moderated, heavy water (D_2O) reflected
- Maximum neutron flux (5×10^{14} n/cm^2.s) sufficient to accommodate various utilization requirements

After examining the similar research reactors under construction abroad – 14 MW ORPHEE in France, 20 MW JRR-3 in Japan, 30 MW JANUS-30 in Indonesia, and 25 MW MAPLE-X in Canada – a decision was made to make a strategic alliance with the AECL for a joint design approach. Basic design was initiated in 1985 to send a KAERI design team to AECL's Whiteshell Nuclear Laboratory in Manitoba and Montreal in Quebec to work under a joint design approach. Subsequently, AECL played both advisor's role and reactor components supplier's role, while the Korean entities assumed the design and construction work under KAERI's project management. KMRR was given a new proper name, HANARO (High-flux Advanced Neutron Application Reactor), upon commissioning. On February 8, 1995, HANARO reached its initial criticality. It was a historic day to remember. Korean scientists and engineers managed to design and construct a state-of-the-art research reactor for the first time. They had overcome the difficulties encountered with project delays and budget overruns. A commemorative stamp was issued and a stone monument was unveiled, with President Kim Yong-sam presiding over the ceremony.[3]

Two 10 MW MAPLE-X reactors were being built in 2002 at Chalk River Nuclear Laboratory to replace the ailing NRX reactor dedicated for radioisotope production. The original 25 MW MAPLE-X reactor was selected to be the reference reactor for HANARO due to their technical affinity and construction schedule. The initial plan for MAPLE-X was to be in operation by 1989, one year ahead of HANARO's original plan. However, two Canadian reactors experienced one of the most unusual circumstances involving legal litigations between the regulatory licensing body and AECL, which resulted in the permanent shutdown of the finished MAPLEs.

Jordan and beyond

If you are not a research reactor specialist, it may be difficult to appreciate the coming of HANARO. A number of design, analysis, and verification tests had to be carried out on critical reactor components in order to satisfy the licensing requirements, which were not always clear-cut. Only minimum utilization facilities, such as neutron radiography and isotope production ports, were installed when HANARO became operational in 1995. Only the basic structural provisions were made for future expansion of users group. Today the reactor hall is full of utilization equipment and scientific instruments gradually installed during the last fifteen years of operation. The latest additions were the fuel test loop and the cold neutron beam port facility, which delivers cold neutron through guide tubes to the new building. One example of HANARO utilization to date was the material irradiation testing and examination on power reactor fuel and structural material, which made critical contributions to the life extension of Kori Unit 1 for ten more years. Accumulated design and operation experience gave system designers at KAERI sufficient confidence to challenge themselves to come up with a new design for a research reactor.

The first opportunity to participate in a foreign research reactor project came in 2009. Together with KOPEC, the KAERI team was awarded a research reactor renovation project at the Greek Demokritos Institute (GRR-1) near Athens to increase power level to 10 MW. They learned how to become competitive in an international bidding situation, competing with better-known reactor vendors like Areva of France and INVAP of Argentina. Soon, other opportunities followed with the international tender of the 80 MW PALLAS reactor at Patten, Netherlands, and 5 MW Jordan Research and Training Reactor (JRTR) near Amman, competing among the same three vendors. While the outcome of the PALLAS bidding is still pending as of January 2011, the supply contract was signed between the Jordan Atomic Energy Commission and KAERI/Daewoo consortium in March 2010 to supply a new 5 MW research reactor to Jordan. The project encompasses the supply of the reactor, together with an isotope production facility and nuclear training center, by 2015 at the Jordan University of Science & Technology (JUST) campus, located seventy kilometers north of Amman. Site preparation work is underway with the PSAR submittal to the Jordanian Nuclear Regulatory Commission planned in 2011. Since its establishment in 1986, JUST has been at the forefront of institutions of higher learning in the Arab world. Together with the University of Jordan in Amman, they are considered to be the top two premier institutions of higher education in engineering and medicine. Jordan has clear intention to move into nuclear power with the establishment of a nuclear engineering department at JUST in 2007. They took the research reactor as the next logical step to nuclear power, as South Korea did in the 1970s.[4]

The year 2009 will be remembered as an epoch-making one in Korean nuclear history for embarking on the journey into the new world of nuclear exports, with a power reactor to the UAE and research reactor to Jordan. An analogy between the two events goes beyond what was happening in the same year and in the same Middle Eastern region. Although the research reactor project scale is much smaller than that of a power reactor, striking similarities exist with how the first units were introduced to Korea under turnkey basis, how technical self-reliance was built up in time, and how they challenged the world majors in winning the international bidding. And, most of all, both share the common root: KAERI. Due recognition came from the National Academy of Engineering of Korea (NAEK) on December 16, 2010, on the special occasion of the "Nation's 100 Most Significant Engineering Achievements in the last 100 years" award ceremony held in Seoul. Three Awards of Excellence were presented to the nuclear sector for the following technical achievements and their respective main players:

<u>Research reactor design, construction, and operation technology</u> (1990s)
  Kim Seong-yun, Oh Se-ki, Ha Jae-joo (KAERI)

<u>Korean Standard Nuclear Power Plant design technology</u> (1990s)
Kim Byung-koo, Lee Byung-ryung, Han Pil-soon (KAERI)
<u>Advanced Power Reactor design and construction technology</u> (2000s)
Rieh Chong-hun, Choi Yeong-sang (KEPCO)

It was a special occasion and honor from the esteemed Academy of Engineering to salute the landmark indigenous technical breakthroughs over the last century that brought Korea out of poverty to experience an economic miracle. Both KAERI and KEPCO's main achievements were awarded simultaneously. Besides being personally honored to receive one of the three awards in the nuclear sector, I had a singular feeling of relief. The peaceful use of nuclear energy was finally recognized by civil society for its role in the economic miracle, overcoming internal and external struggles. Perhaps the long-lasting rivalry and sometimes animosity among the nuclear entities may be coming to an end (or at least moving up to higher level of maturity).

EPILOGUE: ALONG THE *NUCLEAR SILK ROAD*

South Korean success stories are generally credited to its people. They are known to have enviable work ethics, resolute dedication, and unwavering focus on excellence when it comes to critical moments. The 2002 World Cup FIFA game can illustrate this point. The South Korean team, although they were hosting the game together with Japan, was not expected to be a factor in the competition. They never had a single win in the World Cup prior to 2002. As a surprise to everyone, the "Red Devils" made it to the semi-finals, among the very top four countries. My ninety-nine year-old grandmother, who never had anything to do with football in her entire life, stood up all night to cheer for the Red Devils. South Korea defied the world, defied the critics. They brought a new sense of confidence to the nation as never before.

Seven years later, South Koreans defied the world one more time in the nuclear power business; this time the venue was Abu Dhabi on the Gulf Coast. UAE has ambitious national plans to prepare for the post-oil era by selecting nuclear power for their future energy option. The world's best-known nuclear suppliers were in the competition, including the least-known South Korean consortium, headed by KEPCO. As a surprise to everyone, South Korea won the competition based on the assurance of delivery and the local infrastructure support, including the joint operation of the NPPs, besides the basics like safety and economy. This single episode gave me a compelling reason to write this book - to bring the Korean nuclear story to the world. My hope is to demystify the inside stories for international readers and disclose how it happened in South Korea over the past half-century of nuclear power development.

I tried to deliver a correct and complete picture of the Korean nuclear history to the best of my knowledge.

South Korea is the single country that came out of the Least Developed Country (LDC, a UN nomenclature for the poorest country) group to a Development Assistance Committee (DAC, an OECD nomenclature for the donor) country in the past half-century. The nuclear power program vigorously pursued in this small but overpopulated developing country must have played a significant role in elevating the statehood by supplying an ample, reliable, and most of all, affordable source of electric energy for the dynamic engine of growth and development.

Two schools of thought dominated the Korean nuclear power program over the past half-century. One was the research minds that were temporarily turned into the drivers for commercial nuclear power core technology from KAERI; the other was the industrial management minds transformed into competitive nuclear operators and vendors from KEPCO. These two entities had their share of constant internal struggles between localization versus risk-taking; however, credit is due to both entities for providing the brain power and management power at the right times. Starting from a tiny kilowatt research reactor imported from the US fifty years ago, a world-class nuclear industry dynamo is thriving in Korea today. Seemingly a "miracle" story could be better understood with all the internal agonies and infighting that went on behind the scene of every milestone decision, as told in the chapters of this book. South Korea had the right leaders at the right time, from the very top of the government down to the working levels. The 1980s and 1990s, following the Chernobyl accident, were a nuclear dark age in the Western world; however, this period provided an invaluable window of opportunity to build-up the national nuclear power base infrastructure, just in time for the nuclear renaissance. A classical case of turning crisis into opportunity seemed to have worked in Korea. Looking back on the road map to the Koreanization of nuclear power technology as described in this book, we must admit the approach taken was to piggy-back on imported technology by replicating the know-hows and know-whys from the US. Korean engineers were good and fast learners, but should not forget the fact that they had the best mentors from the US. However, a relative lack of depth in basic nuclear sciences and rising complacency could be haunting the Koreans as they move away from the replication technology into an advanced creative arena, where no palatable mentors can be found. Strengthening of international cooperation through joint research projects could be the only way out to share the advanced technologies among like-minded countries. Other developing countries with nuclear power ambitions could benefit from the Korean lessons learned hard way.

The two Koreas had a similar start to their nuclear program until the mid-1970s, when the North took the weapons option while the South chose to focus only on the peaceful use, primarily in nuclear power. Nowhere on this planet are the double faces of the atom more vivid than on the Korean peninsula today. Night-time satellite photos taken over the peninsula testify to the outcome: the northern half is in complete darkness and isolation, the southern half is bright as daylight with openness, largely coming from the twenty NPPs operating base loads at night. This book only covers what happened in the South without mentioning much about the North. The difficulty was to acquire credible reference data on nuclear technology in the North, as they remain the most isolated and closed country in the world, more so in the nuclear area being in total secrecy. The only exception was the KEDO LWR project since the South was trying to replicate one of their KSNP units in the North. For South Koreans, it had more significance than building another NPP as it was deemed a road to reunification. But an unfortunate course of events in the mid-2000s dictated the LWR project to be mothballed. However, the immediate outcome had a positive spin in that the LWR construction experience was readily applied to the first export project success. KEPCO has transformed itself from a large nuclear electric utility company into a nuclear export consortium leader to developing countries. The KEPCO model could shed new light on the export business model, where a fresh approach is being taken as to how to build, operate, and owning NPPs.

The ancient Silk Road was a crossroads of civilization from Asia to Europe (and vice versa) over a millennium, forming the commercial and cultural bonds of Eurasia. Brave merchants and adventurers risked their lives to travel the long journey, to cross the Pamir mountains and Gobi deserts, in caravans of camels and horses on the overland routes. Big sailing ships breezed though the Red Sea, Persian Gulf, Indian Ocean, then through the Strait of Malacca to reach China, Japan and Korea along the coastal routes. Not only were silks and glassware traded, but also technologies like gunpowder were transferred. When UAE signed the contract with the South Korean consortium (and Turkey is negotiating a new NPP with Japan), suddenly the old Silk Road is coming back to life with the building of new nuclear plants in the twenty-first century. Japan and Korea had already built up their fleet of NPPs from indigenous nuclear power technologies, delivering the major bulk of domestic electrical energy. China and India have the most ambitious nuclear power program of any other places in the world today, testifying the coming of the nuclear renaissance in a big way. China is all out for transferring PWR technologies from the US and France to set up their own nuclear power technology infrastructures, as Japan and South Korea did several decades ago. The only difference is that their scale is bigger and the pace is faster. One hundred GWe of new nuclear capacity is said to be on the line in China alone by 2030. India is following suit, not far behind, encouraged by the recent easing of nuclear trade barriers from the US (an exceptional measure to a non-signatory of the NPT). Additional newcomer countries may soon include Iran, UAE, Turkey, Vietnam, Indonesia, Malaysia, and Bangladesh in Asia. Not all have enormous financial wealth, but they all certainly have the national appetite for nuclear power and a thriving economy to back it up. If one plots all the announced NPP sites of these countries superimposed with the old Eurasian trade routes, the new *Nuclear Silk Road* can be visualized. Today, the Emirates Airline makes daily flights from Seoul to Dubai in an A380 super-jumbo carrying some five hundred passengers per plane. Soon enough the container freight ships will sail from Korean ports to Braka NPP site in the UAE along the Silk Road sea routes. (see "Nuclear Silk Road" map in page iii.)

The message is clear. The nuclear renaissance is coming in a big way in Asia in the old Silk Road countries. The new market is big enough for all the major nuclear vendors of today. Interconnectivity among the newcomers and the supplying countries will need to be strengthened further beyond the commercial boundaries to assure absolute safety and security of all operating NPPs in the region. Keeping in mind that just one accident anywhere means the same accident everywhere, international cooperation is needed more than ever before. The true meaning of the "Nuclear Silk Road" is to bring peace and prosperity with safe and economic nuclear power in the years to come.

EPILOGUE:
POST-FUKUSHIMA SECOND THOUGHTS

Japan is considered by many to be one of the safest countries in which to work, live, and travel, from its highly sophisticated civil infrastructure and safety culture imbedded in its everyday life. Public awareness and preparedness against natural disasters like earthquake and tsunami must be the highest in Japan for its long history of seismic activities in the islands. All its fifty-four nuclear power plants and fuel cycle facilities operating in Japan are designed with the most stringent seismic codes and standards established by the world nuclear community. Being the only country which had experienced the radiation hazards from nuclear bomb blasts of some sixty-five years ago, Japanese people are perhaps the most sensitive and best educated about radioactivity released to humans and the environment. Despite all the natural and man-made handicaps against nuclear energy, Japan has chosen to pursue the peaceful uses of nuclear power since the "Atoms for Peace" of 1953 as its national policy. Today, Japan ranks third in the world in generating electricity from nuclear power plants, next to the US and France. Its fleet of NPPs scattered all around its coastlines, is delivering nearly 30% of its electricity needs. Japan's nuclear technology base, among the best in the world, is supported by academia, nuclear research labs, to the industry. Japan has mainly adopted the technology transferred from the US for commercial power reactors over a half-century ago.

Nine electric utility companies operate fifty-four NPPs in Japan. Thirty units are of BWR type located in the northeastern part, and remaining twenty-four units are of PWR type mainly located in the southwestern part. Tokyo Electric Power Company (TEPCO) is the largest privately owned electric utility company in the world with seventeen BWR units in operation; six units at Fukushima Daiichi (meaning the 'first'), four units at Fukushima Daini (the 'second'), and seven units at Kashiwazaki-Kariwa on the west coast. TEPCO has a unique policy of supporting two local nuclear vendor companies, Toshiba and Hitachi, to design and build complete BWR units on an alternating basis. Also the TEPCO power grid uses 50 Hz frequency in the northeast while the southwestern half of Japan uses 60 Hz, which makes the sharing of electricity extremely difficult in emergency situations.

I was about to finish writing this book in early March of 2011 when the fatal earthquake, The Big One, hit the northeastern shore of Japan. Unfolding crises at Fukushima are still continuing as this book is going to print. My intention in adding this extra, unplanned Epilogue is not to give a detailed account and analysis of the accident itself. Rather, I intend to provide a brief overview of the accident and examine what if it had happened to the Korean NPPs. Invaluable lessons to be learned will appear in due course, as it did with the TMI and Chernobyl accidents of bygone years. My keen interest is how it would impact the future of nuclear renaissance, particularly in Asian countries, as I tried to describe the coming of the Nuclear Silk Road in this writing. Will this crisis bring yet another opportunity for safer nuclear powers in the long run?

The Big One

People living in the Pacific Rim, namely in California, Chile, and Japan are much used to earthquakes and tsunamis imbedded in their life-style as a result of active tectonic movements in the area. Sometimes "The Big One" is used to casually describe a mega-thrust earthquake expected to happen in California or Japan with an interval of approximately seventy years or so. On the fateful afternoon of March 11, 2011, such a Big One did come in the northeastern part, Tohoku, of Japan with a magnitude of 9.0 on the Richter scale. It was the fourth largest earthquake in recorded history, only surpassed by the Chilean earthquake of 9.5 magnitude in 1960, the Alaskan of 9.2 in 1964, and the Sumatran of 9.1 in 2004, all along the Pacific Rim. It was the single largest devastating one in Japan, to be known as the "Great Tohoku Earthquake of 2011", coupled with the record high tsunami waves hitting the northeast coastline of Tohoku region. Some described the epic accident brought a calamity of Biblical proportions. Even for the well-prepared civil disaster infrastructure of Japan for such events, it was simply too big and overwhelming in magnitude and scale.

Altogether eleven NPPs were in full operation. All of them went into immediate safe shut-down mode, as they should, right after the quake in the Tohoku area. They were three units at Onagawa, three units at Fukushima Daiichi, four units at Fukushima Daini, and one unit at Tokai, all of BWR types. When the tsunami arrived the coast about forty-five minutes later, the Fukushima Daiichi site experienced the highest tsunami wave (reported to be fourteen meters high), resulting in the total station blackout (SBO), then the subsequent nuclear crises unfolded. The Fukushima units were known to be designed for a six-meter tsunami, and the wave-height was more than twice the design limit.

Fateful Fukushima-First (Daiichi)

Every nuclear power plant is designed to withstand man-made accidents or natural catastrophes to minimize the probability of core damage frequency (or possibility of core meltdown). Considerations of design basis earthquake and tsunami are among the principal design requirements in contrast to any other industry. Should there be any accidents or catastrophes, the reactor should shut down immediately to stop the fission process, with the residual heat removal system automatically taking over to cool down the remaining heat (or residual decay heat) from the fission products. This can only be assured, however, when the cooling water is continuously injected to the reactor vessel to protect the core from melting, as well as maintaining continuous cooling to the spent fuel stored in a pool just outside of the containment building.

Alas, this simple enough first rule of the reactor safety seems to have been severely compromised at the fateful units 1,2,3,4 of Fukushima Daiichi, not because of the record earthquake of 9.0, but because of the fourteen-meter high tsunami. The awesome power of nature disabled all the redundant emergency power systems with one blow beyond anyone's imagination, when the station blackout came after some eight hours of battery life. The defence-in-depth concept was brutally challenged by the common-mode failure. Lack of continuous cooling water with no electricity to pump water into the reactor core and the spent fuel pool resulted in the fuel damage in the reactor vessels and the spent fuel pool. Under these extreme conditions, drying out of uncovered nuclear fuel led to hydrogen production (a well-known phenomenon called "zirconium oxidation") that triggered hydrogen explosions, and spill-over of radioactive materials into the environment. The situation was reported to have reached the International Nuclear and Radiological Event Scale (INES) Level 5 or 6 based on primarily the radiation release to people and the environment. This was revised to Level 7 by the Japanese authority one month later. The IAEA determined the Chernobyl accident of 1986 to be INES Level 7, while the TMI accident of 1979 was Level 5. Both accidents were caused by human errors and equipment malfunctions, thus different in nature from Fukushima case where the cause was natural disaster far beyond the design bases. It is believed to have been further aggravated by the disruption of communication links between the plant and off site emergency response centers. So the question is, how to prevent from such disasters and still protect the core?

What appears to be the single most important key mechanism which initiated the avalanche of subsequent crises at Fukushima was the complete loss of on-site alternating current electric power, known as the "station blackout (SBO)." The tsunami attack disabled the multiple off-site and on-site electric powers, such as emergency diesel generators. If the units' diesel generators were somehow protected from the deadly seawater drowning, the nuclear crisis could have been avoided, as demonstrated by the same BWR units at Onagawa and Fukushima Daini stations on the same Tohoku coastline on that day (Onagawa was even closer to the epicenter). These two sites were on higher elevations to be saved from the tsunami, which made a world of difference. The probability of reaching a complete SBO was considered highly unlikely until now, but taken seriously by the NPP vendors and owners to have it designed into a Beyond Design Basis Event, as recommended by the USNRC Regulatory Guide since 1988.

PWR vs. BWR

NPPs in commercial operation today are dominated by one type of reactor, i.e., Pressurized Water Reactor (PWR); 269 NPPs among a total of 442 units in 30 countries are PWRs (61% by units, 66% by output). When you consider ones under construction at present, this dominance goes over 70% globally. BWRs are operated only in the US, Japan and a few other countries. South Korea was extremely fortunate to have selected PWR reactors from the very beginning, as the standardized flag carrier from Kori Unit 1, and KSNP to APR1400 models.

It is thus important to understand the basic inherent safety differences between BWR and PWR type power reactors. BWR design is based on the single loop connecting the reactor, which generates heat; and the turbine, which generates electricity, by allowing the coolant water to boil in the reactor vessel (thus, the name "Boiling Water"). The advantages are to gain compact design through simplicity and better efficiency. However, the inherent shortcomings are in the radiation contamination and control, particularly, in emergency situations. On the contrary, PWR design is based on dual loops; the primary loop to generate heat in the reactor vessel under high pressure (thus, the name "Pressurized Water") to avoid boiling, and the secondary loop where water is boiled in the heat exchangers called "steam generators" to supply steam to the turbine. This provides one additional radiation barrier in the primary side.

One other important difference is the design concept of the containment building which surrounds the reactor vessel only (for BWR), and the entire primary loop (for PWR). Thus the total containment building cavity volume is up to ten-times greater in PWR than BWR due to the basic layout. Therefore, the mitigation of severe accident becomes more effective in PWR, such as hydrogen explosion which could compromise the breach of containment. After all, the containment building (1.2-meter-thick heavy concrete dome building in all PWRs) is to provide the final radiation barrier to the environment. The case of the TMI (PWR type NPP) accident in 1979 illustrated this point loud and clear. While the reactor itself experienced similar core meltdown resulting from the severe accident as in Fukushima case, radioactivity release to the surrounding environment was kept at a negligible level to protect the public. Ironically, TMI stands as a testimony of ultimate reactor safety, a proven record of radiation protection to the biosphere from the worst severe accident in a PWR nuclear power plant, despite the permanent shutdown of the unit and huge financial loss to the electric company. Lessons learned from TMI for human and equipment errors, and from Fukushima for natural causes, with a large number of safety upgrades in modern NPPs (BWR as well as PWR-types), will make a severe accident even more unlikely.

Accident anyplace, accident everyplace

Fukushima impacted on the world's nuclear industry as if it had happened to every nuclear power country. The common bondage of nuclear safety issue from any NPP is being reverberated in every NPP in the world. Only difference this time was the fact that the giant force of nature which was over and beyond comprehension, not the human error, initiated the accident. The enviable track records of the global nuclear safety over a quarter-of-a-century since Chernobyl are coming suddenly under public scrutiny. Global nuclear industry as a whole will undoubtedly go through the lessons-to-be-learned follow-up actions from the Fukushima accident in the years to come, judging from the past experiences of TMI and Chernobyl. Most nuclear power countries have already committed themselves to review safety arrangements of their nuclear fleets, or "stress tests", on the basis of the Fukushima accident as they become known, cross-cutting reactor types and fuel cycle facilities. Clearly, we need to look again and ensure that every NPP in every country has reliable post-shutdown cooling capabilities even after the most extreme and unlikely natural events.

On a technical level, the nuclear industry must and will meet safety challenges on a new plateau: to survive combinations of extraordinary scenarios beyond their design basis events, including natural disasters, terrorist attacks, and as-yet unimagined disasters and difficulties in their locality. Advanced countries with their own nuclear power technology bases will spearhead the new enhanced safety upgrades and back-fittings as deemed necessary. The IAEA's role in nuclear safety is expected to be strengthened to coordinate and improve the safety standards, and to enhance peer review services to emerging nuclear power countries. Perhaps the time is ripe to consider strengthening the safety standards to be more legally binding, than in the voluntary mechanism as is the case today.

Preliminary thoughts on the follow-up actions in technical aspects for global nuclear industry may include;

1. Enhanced protection against the station blackout,
2. Strengthened spent fuel pool cooling, and spent fuel management,
3. Effective hydrogen removal system,
4. Special attention to the aged NPPs with strengthened life-extension criteria,
5. Adoption of passive safety features for residual heat removal,
6. Applied research in severe accidents, reflected in NPP design/analysis, and emergency manuals,
7. More stringent requirements for NPP sites,
8. Enhanced Emergency Response Facilities.

Additional actions in management aspects may also include;
1. Emergency preparedness, crucial decision-making process at the control tower, particularly in prevention and mitigation of severe accident scenarios,
2. Strengthening the electric utility company's corporate safety culture under extreme conditions together with national policy backup,
3. Reassessment of the independence of regulatory infrastructure,
4. Enhanced information sharing under emergency situations, especially among regional NPP counties (for example, Japanese/Chinese/Korean regulators and operators),
5. Strengthened role of the IAEA nuclear safety in raising the safety standards and the peer review capacity.

Great nuclear dilemma

All said and done, the post-Fukushima follow-ups will take years in enhancing nuclear safety to a higher level to protect from extreme conditions, natural or man-made, by incorporating them into design basis events. No doubt it will impact on the cost of nuclear electricity, narrowing the gap between nuclear and other sources of electricity. Public confidence and support for nuclear power was just about to take off with a global nuclear renaissance, particularly in the Asian countries, as depicted in this book. But regaining public acceptance following the Fukushima event could be a bigger and longer challenge as experienced after the TMI and Chernobyl in the past.

Hence, the great nuclear dilemma. The perceived overwhelming risk factor to the general public must be properly addressed before nuclear power can be viable again. For the best nuclear safety, just good planning and good engineering is not enough. We need the sort of society that can produce accountability and transparency, one that can build institutions that receive and deserve trust. At the present, there is no alternate source of electricity that can replace nuclear, in a low-carbon green-growth economy, trying to move away from fossil fuel options. Most renewables may come too short in scale and economy to replace nuclear. This is why modern society cannot turn its back on nuclear power. A timely global debate was hosted by *the Economist.com,* by posting the question *"This house believes that the world would be better off without nuclear power."* one month after the Fukushima accident. *The Economist*, a weekly newspaper from London, has been conducting timely debates on major global issues through the internet platform. After the active participation of world opinion leaders giving pros and cons on the nuclear power arguments, the final vote was "39% Yes, and 61% No."[1] It was a resounding vote of confidence on nuclear power in a global perspective, as it came from one of the most respected international opinion media, with the lessons learned from the recent tragedy in Japan. Simply put, the case for nuclear power has not changed since Fukushima: the need for continuous, reliable supply of affordable electricity, the importance of energy security so that the fuel for electricity generation is not subject to geopolitical uncertainties, and the need to minimize carbon dioxide emissions from power generation.

For South Korea, it is more significant to deliberate on nuclear power options at home and abroad as it has earned the nuclear vendor country status with the recent launching of the export project in the Middle East. The technical self-reliance based on PWR-type nuclear power established in Korea, which is the main theme of this book, is about to provide high-value-added domestic and export economic benefits. Lessons from Fukushima will no doubt strengthen the safety of nuclear power in every country with a higher cost of nuclear electricity. In a longer term perspective, however, slowing down of overambitious nuclear new-build programs in some Asian countries, which is the backbone of the nuclear renaissance, could be productive and beneficial in gaining time. Lessons learned from Fukushima could galvanize the potential to strengthen their safety infrastructures, and perhaps sharpening the safety culture mentality. It may be a true blessing in disguise as we overcome the Fukushima tragedy.

ANNEX I.
CHRONOLOGY OF NUCLEAR POWER TECHNOLOGY IN KOREA

1956. 2. 3	ROK/US Agreement for Cooperation Concerning Civil Uses of Atomic Energy was singed
1956.	American engineer and entrepreneur Walker Cisler briefed President Rhee Syngman on the nuclear power for Korea, stressing the manpower development. The first group of Korean scientists were sent to the US and UK for nuclear training under the government fellowships
1956. 3. 9	Atomic Energy Bureau (AEB) established in the Ministry of Education
1957. 8. 8	ROK joined the IAEA as a founding Member State
1958. 3. 1	Hanyang University opened first Nuclear Engineering Department in Korea
1958. 3. 11	First Atomic Energy Act passed the ROK National Assembly
1959. 1. 21	Office of Atomic Energy (OAE) established to replace the AEB
1959. 3. 1	Atomic Energy Research Institute (AERI) established as a government entity under the OAE; Department of Nuclear Engineering opened first class at Seoul National University
1962. 3. 28	First research reactor, TRIGA Mark-II (KRR-1, 100 kW), inaugurated at AERI
1968. 4. 9	KEPCO designated for NPP construction and operation

1969. 10. 2	First NPP Kori-1 construction plan formally approved by ROK government
1972. 7. 2	Second research reactor, TRIGA Mark-III (KRR-2, 2 MW), inaugurated at AERI
1973. 4. 6	Special non-government entity KAERI founded with a national assembly legislation to succeed AERI
1974.	KAERI started to recruit Korean scientists and engineers with nuclear power background from the US
1975. 3. 20	ROK National Assembly ratified NPT
1975. 10. 1	KAERI established architect-engineering company KABAR jointly with Burns & Roe of the US
1975. 10. 31	ROK/IAEA signed full-scope safeguards agreement (INFCIRC/140)
1975. 11. 14	ROK NPT entered into full force.
1976. 1. 23	KAERI reprocessing pilot project formally abolished
1976. 10. 12	Burns & Roe pulled out of KABAR, restructured to be Korea Nuclear Engineering (KNE), wholly owned by KAERI
1976. 11. 18	ROK government designated KNE to be the sole entity for nuclear architect-engineering localization (with signature of President Park Chung-hee)
1976. 12. 1	Korea Nuclear Fuel Development Institute (KNFDI) branched out from KAERI
1977. 7. 14	First ROK/US annual JSCNOET meeting held in Seoul
1978. 7. 20	First NPP Kori-1 entered into commercial operation
1979. 4. 11	TMI accident in the US
1979. 12. 12	Chun Doo-hwan came into power following the assassination of President Park Chung-hee on October 26, 1979
1980. 4. 9	New KAERI President Cha Jong-hee takes office following death of President Hyun Kyung-ho
1980. 9. 2	Prof. Lee Cheong-o of KAIST appointed to the Minister of Science & Technology

1980. 11. 19	MOST concludes new restructuring plan for national research institutes, including KAERI
1980. 12. 19	KAERI merged with KNFDI, now "A" standing for "Advanced" and not "Atomic"
1982. 2.	Nuclear Safety Center inaugurated under KAERI, with first Director Kim Dong-hoon
1982. 3. 16	Han Pil-soon of ADD appointed to the vice-president in charge of KAERI Daedeok Engineering Center
1982. 7. 6	KNE restructured to be Korea Power Engineering Company (KOPEC) under KEPCO control; Chung Kun-mo became its first president
1982. 11. 11	Korea Nuclear Fuel Company (KNFC) inaugurated as a subsidiary of KEPCO (Kim Sun-chang was named the first president)
1983. 3.	Park Jung-ki, president of KHIC, appointed president of KEPCO
1983. 4. 8	First prototype CANDU fuels fabricated by KAERI were loaded at NRU reactor at Chalk River, Canada
1983. 4. 12	President Chun Doo-hwan visited KAERI, toured CANDU fuel fabrication plant
1983. 5. 31	President Park Jung-ki of KEPCO elected to be the KAERI board chairman
1983. 7. 6	Han Pil-soon appointed as the second president of KNFC
1984. 4. 9	Han Pil-soon inaugurated as the fourth president of KAERI
1984. 8. 30	Brainstorming of KAERI senior staff concluded KAERI's scope in new NPP localization program to be limited to reactor vessel only
1984. 9. 8	First KAERI-made twenty-four CANDU fuel bundles loaded into Wolsong NPP
1984. 9. 18	MER minister Choi Dong-kyu, with the Electric Power Group Cooperation Council presidents, concludes designation of responsibilities for technical self-reliance (KAERI with reactor vessel and core design

	while KOPEC takes the rest of NSSS system design plus overall architect-engineering work)
1984. 10.	NPP Standardization Study for technical self-reliance completed among MER/MOST, KEPCO, KOPEC, KHIC, and KAERI after three years
1985. 2.	ROK government announces firm construction plan for YGN units under the standardized NPP with technical self-reliance program
1985. 6. 25	EPGCC presidents' meeting reverts to the September 1984 decision on designation of responsibilities, this time for all NSSS system design assigned to KAERI
1985. 7. 1	Power Reactor System Division established at KAERI with thirty-three starting members (Kim Byung-koo the first director)
1985. 7. 29	214th Atomic Energy Commission endorses formal blessing to YGN construction plan and assigns KAERI as the responsible entity for NSSS reactor system design and initial core design for future NPP projects
1985. 8. 26	LWR Nuclear Fuel Technology Inducement Contract signed between KAERI/KWU of Germany
1985. 11. 4	Invitation-to-Bid (ITB) for YGN project with technology transfer sent out to twenty-three companies in seven countries for international competitive bidding
1986. 4. 1	Evaluation of the bids started (only four submitted for NSSS) on the system design, technology transfer conditions, and the joint design program
1986. 4. 26	Chernobyl nuclear accident in Ukraine
1986. 7. 10	KAERI submitted the YGN NSSS system/initial core design evaluation results to KEPCO
1986. 9. 30	The YGN successful bidders announced by KEPCO (NSSS: CE, A/E: S&L, T/G: GE)
1986. 10. 13	MOU on advance training for KAERI system designers signed with CE
1986. 12. 14	First KAERI system design team of forty-eight sent to Windsor, USA, for training and Joint Design

1987. 4. 9	Supply and Technology Transfer contracts signed for YGN among all Korean and US entities
1987. 5. 1	KEPCO's Authorization to Proceed on YGN project issued
1987. 5. 16	ROK government approved the YGN Technology Transfer contracts
1987. 6. 19	ROK government approved the YGN Supply contracts
1987. 7. 1	First mass-produced CANDU nuclear fuels from KAERI delivered to KEPCO's Wolsong NPP site
1987. 8. 25-30	First YGN Project Review Meeting held at KEPCO, Seoul
1988. 5. 1	Design center transition plan agreed between CE and KAERI
1988. 5. 30	The second YGN Project Review Meeting held at CE, Windsor
1988. 5. 31	YGN PSAR submitted to MOST, the regulatory body
1988. 10	National Assembly investigation on possible YGN irregularities began
1989. 4. 18	NSSS system design center transferred from Windsor to Daedeok
1989. 12. 21	YGN Construction Permit issued by MOST
1989. 12. 23	First concrete poured at YGN Unit 3 reactor building
1989. 12. 30	KAERI's name with "A" for "Atomic" reinstated from "Advanced"
1990. 2. 15	Nuclear Safety Center became independent from KAERI; Korea Institute of Nuclear Safety (KINS) inaugurated
1990. 9. 6	First technology self-reliance evaluation report issued by KAERI
1991. 7. 2	First KSNP UCN 3&4 units supply contracts signed
1991. 12.	North/South Koreas Joint Declaration of the Denuclearization of the Korean Peninsula was promulgated
1992. 12. 28	YGN FSAR submitted to the regulatory body (MOST/KINS)
1993. 11. 10	YGN Unit 3 Cold Hydro Test started

1994. 9. 9	YGN Operation License issued by MOST; initial fuel loading began
1995. 3. 30	YGN Unit 3 performance warranty test completed
1995. 3. 31	YGN Unit 3 started commercial operation
1995. 12. 7	Technical self-reliance evaluation symposium held at KEPCO, original target of 95 percent reached (copy technology)
1996. 1. 1	YGN Unit 4 started commercial operation
1997. 1. 1	KAERI NPP-related commercial projects and manpower transferred to private sector: reactor system design to KOPEC, initial core design and CANDU fuel fabrication to KNF, and radwaste to KEPCO
1997. 12.	YGN Unit 3 honored with the "Best Nuclear Project of the Year Award" from the International Power Engineering (Park Yong-taek, the YGN project manager, received the award)
1998. 8.	UCN Unit 3 started commercial operation
1999. 12.	UCN Unit 4 started commercial operation
2002. 5. 7	KHNP received the Standard Design Approval for APR1400 model from the Korean regulatory body MOST/KINS
2007. 11. 28	First advanced APR1400 units started construction at Shin-Kori 3&4.
2009. 12. 27	First NPP export contract signed between the UAE ENEC and KEPCO consortium for 4xAPR1400
2010. 3. 30	First research reactor export contract signed between Jordan AEC and KAERI/Daewoo consortium for 5 MW JRTR
2010. 12. 27	Korean government promulgated the 'National Day of Nuclear Safety and Promotion' ('Atomic Energy Day' in short) to replace the 'Nuclear Safety Day' on the first anniversary of the UAE contract
2011. 3. 11	Nuclear accident of INES Level 7 occurred at Fukushima Daiichi NPP in Japan following the Tohoku earthquake and the subsequent tsunami

ANNEX II.
KOREAN NUCLEAR ENTITIES

Many nuclear-related public and private, industrial organizations have emerged and evolved as the nation's nuclear program expanded over the last half-century. As such, their names and affiliations changed from mergers and restructurings in recent years. Over two hundred and sixty nuclear entities are listed in the 2010 Korea Atomic Industrial Forum Directory of Personnel (consult www.kaif.or.kr for more information). The list below summarizes the major nuclear entities today with their missions, overview in their nuclear business, brief history, and English web site information in the following sectors:

Government sector

Atomic Energy Commission (AEC)
Nation's highest policy-making organ in nuclear matters. The prime minister serves as the chairman with ten commissioners, all of whom are non-standing members.

Ministry of Education, Science & Technology (MEST)
Ministry of Education and Ministry of Science & Technology merged into MEST since 2008. Responsible for nuclear policy, international co-operation, nuclear safety & regulatory, and nuclear research.
Web site: www.mest.go.kr

Ministry of Foreign Affairs & Trade (MOFAT)
Ministry of Foreign Affairs and the trade part of the Ministry of Commerce & Industry merged into MOFAT since 2008. Responsible for international nuclear diplomacy in general, nuclear nonproliferation, and disarmament matters in particular.
Web site: www.mofat.go.kr

Ministry of Knowledge Economy (MKE)
Ministry of Energy & Resources and the industry part of the Ministry of Commerce & Industry merged into MKE since 2008. Responsible for energy planning, nuclear industries, and nuclear utilities with radwaste management.
Web site: www.mke.go.kr

Nuclear Safety Commission (NSC)
Highest policy-making body in nuclear safety matters, including regulatory licensing. Minister of MEST is the chairman with seven nonstanding members.

Public/research sector

Daedeok Innopolis (DDI)
Started as Daedeok Science Town, northern suburb of Daejeon (160 km south of Seoul), in the mid-1970s as the nation's research hub in science and technology. Today it encompasses over fifty research centers and two hundred high-tech venture firms administered by the Daedeok Innopolis under MKE. Twelve nuclear-related public research centers and university, plus twenty some nuclear venture service companies at the "Nuclear Valley" are thriving in the DDI, earning it the nickname "Nuclear technology Mecca".
Web site: www.ddi.or.kr

Korea Atomic Energy Research Institute (KAERI)

The oldest national nuclear research center since 1959. Its name changed three times; the original Atomic Energy Research Institute to KAERI in 1973, merged with KNFDI and "Atomic" changed to "Advanced" in 1980, then restored back to "Atomic" in 1989. KAERI's main campus relocated from Seoul to Daedeok in the early 1980s; it served as parent company to many daughter companies, including KOPEC, KNFC, KINS, KRMC, KIRAMS, and KINAC. Responsible for nuclear research in basic science, nuclear safety, advanced reactors, and fuel cycle technologies. Engaged in the research reactor export projects based on the home grown HANARO technology.
Web site: www.kaeri.re.kr

Korea Electric Power Research Institute (KEPRI)

KEPRI's roots go back to the establishment of the Electricity Laboratory Testing Center in 1961, renamed 'KEPRI' in 1980, and relocated to the Daedeok Innopolis in 1993. Its mission is to conduct various research projects in support of KEPCO, including the Nuclear Power Laboratory.
Web site: www.kepri.re.kr

Korea Institute of Nuclear Nonproliferation and Control (KINAC)

KINAC is a logical outgrowth of the Technology Center for Nuclear Control (TCNC) at KAERI since 1994, when the Additional Protocol with the IAEA became effective in 2004. Its missions are to form the State System of Accountancy Control for the national safeguards inspections, together with physical protection and export control activities to address nonproliferation matters in technical fields.
Web site: www.kinac.re.kr

Korea Institute of Nuclear Safety (KINS)

KINS became independent from KAERI in 1990, having originated from the Nuclear Safety Center when the YGN NPPs were under construction, with KAERI participating in the NSSS system design and initial core design projects. Its major missions are to review and license

the new build projects as well as to conduct periodic safety inspections for the operating units. Most active in international cooperation with other national nuclear regulatory agencies, including the reactor safety school at the Daedeok campus.
Web site: www.kins.re.kr

Korea Institute of Radiological & Medical Sciences (KIRAMS)
KIRAMS was a logical outgrowth of the Cancer Center Hospital in Seoul from KAERI's medical research in radiological science applications in 2007. It is one of the leading cancer research hospitals in Korea with medical isotopes from HANARO. They were the first hospital to introduce PET cancer diagnosis system with the indigenous medical cyclotron.
Web site: www.kirams.re.kr

Korea Radioactive waste Management Corporation (KRMC)
KRMC originated from KAERI's Nuclear Environment Technology Center in 1997, when the radwaste project was transferred to KEPCO. With the promulgation of the radwaste management legislation in 2008, KRMC became independent from KEPCO/KHNP to manage the construction of the low/medium radwaste repository at Gyeongju near Wolsong NPP site. Its missions are to manage the transport, storage, treatment, and final disposal of all radioactive wastes generated from NPPs and other sources.
Web site: www.krmc.or.kr

National Fusion Research Institute (NFRI)
NFRI is the nation's leading nuclear fusion research center with operation of the K-STAR (Korea Superconducting Tokamak Advanced Research) reactor since 2008 at the Daedeok Innopolis campus. Also participates in the ITER (International Thermonuclear Experimental Reactor) project being built at Cadarache, France together with China, EU, India, Japan, Russia and USA.
Web site: www.nfri.re.kr

National Research Foundation of Korea (NRF)
NRF was restructured to merge the Korea Science Foundation and Korea Academic Foundation for humanities in 2008, expanding its head office in Daedeok Innopolis. Its mission is to provide research grants to scholars and researchers in science/engineering and humanities/social sciences. Center for Nuclear R&D at NRF funds research projects in nuclear fields to national research labs and universities.
Web site: www.nrf.re.kr

Pohang Accelerator Laboratory (PAL)
PAL is a synchrotron radiation accelerator built and operated at Pohang University of Science & Technology, fully funded by the steel giant POSCO since 1994. Its missions are to conduct basic experimental nuclear research using the beam lines at PAL.
Web site: http://pal.postech.ac.kr

Nuclear industries/utilities sector

AECL-Korea
AECL of Canada has operated its Korean office in Seoul since 1976 in support of the four-unit CANDU stations at Wolsong site.
Web site: www.aecl.ca

AREVA Korea, Ltd.
AREVA of France has operated its Korean office in Seoul since 2003 in support of Ulchin Units 1&2, which were supplied by Framatome, and associated fuel service business in Korea.
Web site: www.areva.com

Daewoo E&C Co., Ltd.
Daewoo Construction has been one of the leading civil construction company participating in the domestic NPP projects, including Shin-Wolsong Units 1&2 and KRMC radwaste depository construction at

Wolsong site. Daewoo formed a consortium in 2010 with KAERI for the construction of a research reactor in Jordan.
Web site: www.dwconst.co.kr

Doosan Heavy Industries & Construction Co., Ltd. (DHI)
Doosan Heavy Industries originated from the Hyundai International days in the 1970s to start the modern heavy industry plants at Changwon; later became Korea Heavy Industries & Construction (KHIC) under government ownership. Component design and manufacturing technologies for PWR reactor plant as well as the turbine/generator plant were set up during the KSNP projects. Privatization to Doosan came in 2000 with thriving nuclear business group in time for the global nuclear renaissance.
Web site: www.doosanheavy.com

Hyundai Engineering & Construction Co., Ltd.
Hyundai Construction is the founding member of the Hyundai Group, started by the legendary founder Chung Ju-yung from the 1950s for shipbuilding and automobile business, creating one of the most dynamic *chaebol* conglomerates in Korea. Hyundai Construction was the civil constructor from Kori Unit 1 NPP project and remained in the leading position for all subsequent nuclear construction projects in Korea, and the first overseas project in UAE.
Web site: www.hdec.co.kr

Hyundai Heavy Industries Co., Ltd. (HHI)
Hyundai Heavy Industries started from a small shipbuilding company in Ulsan from its parent company, Hyundai Construction, in 1972, becoming the largest shipbuilding company in the world today. HHI made its nuclear connection with nuclear-grade welding technology during the Kori Unit 1 construction; later moved into heavy component manufacturing like steam generators.
Web site: www.hhi.co.kr

Korea Electric Power Corporation (KEPCO)

KEPCO originated from the Korea Electric Company (KECO) since 1961, has been the government-owned sole electric utility company responsible for generation, distribution, and sales of electricity in Korea. In 2000 KEPCO was restructured into six regional fossil generation companies, except KHNP for hydro and nuclear generation, creating independent entities. KEPCO remains the holding company of the regionals and KHNP, and retained the leading role of Korean consortium for nuclear export market. KEPCO was the prime contractor to KEDO LWR project, and is the prime contractor to ENEC of UAE for the construction of APR1400 units to Braka site on the Persian Gulf. KEPCO also served as parent company to several daughter companies in the nuclear sector, including KHNP, KOPEC, KNFC, and KPS.
Web site: www.kepco.co.kr

KEPCO-E&C

KEPCO-E&C (Engineering & Construction) was the new name given to KOPEC in 2010 to consolidate the nuclear export efforts with KEPCO at the leading position. KOPEC originated from KABAR and KNE in 1975, from KAERI, then gradually KEPCO took over. Also the NSSS system design projects were transferred from KAERI to KOPEC in 1997 to become the overall design and engineering company for NPP projects at home and abroad. The head office is located in Yongin while the Power Reactor System Design Division is located in Daedeok Innopolis within the KAERI campus.
Web site: www.kepco-enc.com

KEPCO-NF

KEPCO-NF changed its name from KNF (Korea Nuclear Fuel) in 2010 together with KEPCO-E&C. It is the sole nuclear fuel fabricator for all PWR and PHWR NPPs in Korea, co-located at the KAERI campus in Daedeok Innopolis. KNF was originated as a daughter company of KAERI and KEPCO in 1981, then the fuel design and PHWR fabrication team were transferred from KAERI in 1997. Zirconium fuel cladding plant was added to KNF in 2009 for full production. As a part

of the KEPCO consortium, KEPCO-NF will be responsible for fuel supply to future export projects.
Web site: www.knfc.co.kr

Korea Plant Service & Engineering Co., Ltd. (KPS)
KPS was created in 1984 as a wholly owned subsidiary of KEPCO, specializing in the maintenance and repair for all power plants in Korea, including nuclear power plants. It operates KPS site offices at all four nuclear sites plus the maintenance technology center at the Kori Training Center.
Web site: www.kps.co.kr

Korea Hydro & Nuclear Power Co., Ltd. (KHNP)
KHNP was separated from KEPCO in 2000; responsible for all construction, operation, and maintenance of NPPs and hydro electric plants, becoming one of the largest nuclear utilities in the world. Its head office is in Seoul; however, large site operations are dominating the KHNP framework at four nuclear sites: Kori, Wolsong, Yonggwang, and Ulchin. In addition, Nuclear Engineering and Technology Center (KHNP-NETEC) is located at Daedeok Innopolis, responsible for the next-generation reactor design development including the APR1400.
Web site: www.khnp.co.kr

Samsung C&T Corporation - Construction
As a member of Samsung Group companies, Samsung Construction is a relatively new emerging construction firm in Korea with Shin-Wolsong 1&2 project and the KRMC radwaste repository construction at Wolsong. With its reputation of building the Khalifa Tower in Dubai (the tallest building on earth), Samsung is participating in the UAE NPP construction.
Web site: www.secc.co.kr

Westinghouse Korea
Westinghouse Korea has a long history of operation since the Kori Unit 1 days in the 1970s as the nuclear vendor to six early NPPs at Kori and

Yonggwang sites. Westinghouse merged with Combustion Engineering in 2000, who was then the supplier of the System 80 technology for the fleet of KSNPs built in Korea, and took over the CE scope of work. As such, Westinghouse today supplies selected equipment to domestic and overseas projects.
Web site: www.westinghousenuclear.com

Academic societies and associations sector

Korea Electric Association – KEPIC Division
Korea Electric Power Industry Code (KEPIC) division of KEA is responsible for instituting industry codes and standards for nuclear components and systems by integrating mechanical codes (ASME), civil/structural codes (ACI), electrical codes (IEEE), and nuclear codes (NEMA) since 1995. Over 220 local suppliers are certified by KEPIC to form the supply chain of qualified vendors for all NPP projects.
Web site: www.kepic.or.kr

Korea Radioisotope Association (KRIA)
KRIA is the coordinating association for radioisotope production, import, transportation, and distribution in Korea since 1985 for medical and industrial applications of radioisotopes.
Web site: www.ri.or.kr

Korean Association for Radiation Protection (KARP)
KARP is the academic society to promote radiation safety and protection for radiation workers and the general public since 1975; publishes the *Journal of Radiation Protection*. It also serves as the Korean counterpart to the International Commission on Radiological Protection (ICRP), which formulates the international standards on radiation protection.
Web site: www.karp.or.kr

Korean Society for Nondestructive Testing (KSNT)
KSNT is the academic society to promote technologies in nondestructive testing for industries. It has wide applications in nuclear power plants during construction and operation. It also provides education and training for the ASNT Level III certification program to the pre-service and in-service inspections for NPPs.
Web site: www.ksnt.or.kr

Women in Nuclear - Korea (WiNK)
WiN-Korea was established in 2003 in conjunction with the WiN-Global movement which started from Europe in 1993 in order to generate public acceptance of nuclear energy from women's participation. It also serves as a forum of sharing information among female nuclear professionals in medical and technical fields at home and abroad.
Web site: www.winkorea.or.kr

Korea Nuclear International Cooperation Foundation (KONICOF)
KONICOF was created in 2004 from the Nuclear International Cooperation activities of the MEST to strengthen bilateral cooperation agreements and multilateral agreements with IAEA and OECD-NEA, with its office at Daedeok Innopolis inside KAERI campus.
Web site: www.konicof.or.kr

Korea Nuclear Energy Foundation (KNEF or Korea Nuclear Energy Promotion Agency)
KNEF was created in 1992 by the Korean government as an independent entity to promote public acceptance of nuclear energy by providing credible information to the general public. Its main function is to organize education and training seminars, nuclear site tours for the general public, and conduct public opinion surveys.
Web site: www.knef.or.kr, or www.konepa.or.kr

Korea Atomic Industrial Forum (KAIF)
KAIF has provided a unique forum of information exchange among nuclear industries in Korea since 1972, similar to AIF of the US and JAIF

of Japan. Together with the Korean Nuclear Society, KAIF conducts the largest annual nuclear convention in Korea every April. Also hosts international meetings and seminars, publishes books and the bimonthly journal, *Nuclear Industry*.
Web site: www.kaif.or.kr

Korean Nuclear Society (KNS)
KNS was created in 1969 as an academic society of nuclear science and technology, with its publication of journals, including the SCI-listed *Nuclear Engineering and Technology*, published bimonthly in English. Also conducts biannual KNS conventions and a number of research working group committees in technical subject areas. KNS head office is located at Daedeok Innopolis in Daejeon, where a number of nuclear technical entities are co-located.
Web site: www.nuclear.or.kr

University sector (with nuclear engineering departments)

Chosun University, Gwangju
Department of Nuclear Engineering was established at Chosun University in 1985; located in Gwangju, South Cholla Province, with undergraduate and graduate curricula.
Web site: www.chosun.ac.kr/~nuclear

Dongguk University, Gyeongju
College of Energy & Environment was established at the Gyeongju campus since 2008, offering undergraduate programs in nuclear and energy engineering options.
Web site: www.energy.dongguk.ac.kr

Hanyang University, Seoul
Department of Nuclear Engineering at Hanyang University was created in 1958, being the very first nuclear engineering department in Korea. Since the beginning, it educated and supplied the bulk of qualified

engineers needed for the expanding Korean nuclear program. Its main campus with undergraduate and graduate programs is located in Seoul.
Web site: http://nuclear.hanyang.ac.kr

Jeju National University, Jeju
Department of Nuclear & Energy Engineering was established in 1984 to offer majors in mechanical engineering, energy systems engineering including nuclear energy option, and mechatronics.
Web site: http://energy.cheju.ac.kr

KEPCO International Nuclear Graduate School (K-INGS), Kori
Newly established graduate school for specialized training on nuclear power technology, focusing on power reactor systems engineering to be open in 2012. International and Korean students (50/50) for MS program (in English) are recruited at its new campus located at Shin-Kori site.
Web site: www.k-ings.ac.kr

Korea Advanced Institute of Science & Technology (KAIST), Daejeon
Department of Nuclear & Quantum Engineering was first established in Seoul for graduate program in 1981, then relocated to Daedeok Innopolis campus in 1990, with addition of the undergraduate program. Today KAIST is among the top one hundred universities in the world for its academic excellence. Since 2010, KAIST is supporting the Khalifa University in Abu Dhabi, to establish a new nuclear engineering department as part of the UAE's national NPP construction program.
Web site: http://nuclear.kaist.ac.kr

Kyung Hee University, Suwon
Department of Nuclear Engineering was established at Kyung Hee University Suwon campus in 1979 with the introduction of an experimental critical assembly (AGN-201 reactor); now offering undergraduate and graduate program.
Web site: http://ne.khu.ac.kr

Seoul National University, Seoul
Seoul National University (SNU) opened the Department of Nuclear Engineering at its College of Engineering in 1959, the same year KAERI was established immediately adjacent to the SNU, College of Engineering campus in Seoul. Being the nation's highest university, it attracted the top-ranked students in the early years. Today SNU is among the top one hundred universities in the world for its academic excellence, together with KAIST and Postech.
Web site: http://nucleng.snu.ac.kr

Ulsan National Institute of Science & Technology (UNIST), Ulsan
UNIST has established the Interdisciplinary School of Green Energy with nuclear track since 2009, emphasizing the development of advanced reactor systems and safety in nuclear power plants, benefiting from its proximity to the nation's NPP sites.
Web site: www.unist.ac.kr

NOTES

Chapter 1

1. Author interview with Kang Bak-kwang.
2. Kang Bak-kwang, et al., "A study on the effects of '70s-'90s major science policies on S&T and industrial development," MOST Policy Research 2006-21 publication, pp. 3-8, 2007.
3. Yoon Young-ku, "Recollection of My KAERI Presidency (1971.8 ~ 1978.3) – KAERI's Infrastructure and International Hardship in the 1970s," *Treatise in Commemoration of Prof. Yoon Young-ku's Hwan-gap*, KAIST Nuclear Engineering Department, pp. 347-357, 1989.
4. Cha Jong-hee, et al., *Heavy Water Power Reactor Feasibility Study: Canada travel report*, KAERI/PP-74/16, 1973.
5. Lee Hae, George Pon, et al., *Korean CANDU Program: Joint Canada Korea Study*, KAERI/AECL publication, June 1978.
6. Lee Hae, *Establishment of Nuclear Engineering Test & Evaluation Center*, KAERI/421/RR-154/80, 1980.
7. KNFDI, *CANDU-PHWR Nuclear Fuel Localization Plan*, internal report, July 1980.
8. KNFDI, *PWR Nuclear Fuel Localization Project Plan*, internal report, July 1980.
9. Author interview with Nam Jang-soo.

Chapter 2

1. Author interview with Kang Bak-kwang.
2. Author interview with Nam Jang-soo.
3. Cha Jong-hee, *Era of Glory and Indulgence – 30 Years in Nuclear* (in Korean), Shinwoo-sa, 1994, pp. 216-217.
4. Author interview with Ho Nam.
5. Author interview with Cha Jong-hee.

6. Author e-mail communication with Phil Colton.
7. Author interview with Han Pil-soon.

Chapter 3
1. Author interview with Han Pil-soon.
2. Author interview with Chang In-soon.
3. Author interview with Suh Suk-chun.
4. Author interview with Han Pil-soon.
5. Author interview with Kim Sun-chang.
6. Kim Dong-hoon, et al., *30 Year History of KAERI*, pp. 468-472, 1989.
7. Han Pil-soon, "Inauguration Speech as KAERI President", April 9, 1984.
8. Author interview with Han Pil-soon.
9. Author interview with Park Jung-ki.

Chapter 4
1. Kim Jong-shin, et al., *30 Year History of Nuclear Power* (in Korean), KHNP, Vol. 1, pp. 108-111, 2008.
2. Author interview with Shim Chang-saeng.
3. Han Pil-soon, *From hand grenade to reactor*, sections (70) – (73).
4. Author interview with Park Jung-ki.
5. "Meeting minutes of the 4th EPGCC main meeting, 25 June 1985, KEPCO internal document.

Chapter 5
1. "YGN NSSS Joint Design Bid Evaluation Report", KAERI internal report, June 30, 1986.
2. "YGN NSSS System Evaluation Report", KAERI internal report, June 30, 1986.
3. "YGN NSSS Technology Transfer Bid Evaluation Report", KAERI internal report, June 30, 1986.
4. Author interview with Lee Chang-kun.
5. Author interview with Park Jung-ki.
6. Author interview with Kang Bak-kwang.

7. Han Ki-in, et al., *Final report – Yonggwang 3&4 reactor system design project* (in Korean) KAERI, 1995.

Chapter 7
1. Chang Sun-sup, et al., *White Paper on LWR Project* (in Korean), Office of Planning for the LWR Project, 2007.

Chapter 8
1. Syngman Rhee, *Neutrality As Influenced by the United States*, (PhD Dissertation, Princeton University, 1912), Kwanaksa, Seoul, 2004.
2. Dong-Won Kim, "Imaginary Savior: The Image of the Nuclear Bomb in Korea, 1945-1960," *Historia Scientiarum*, 19-2 (2009).
3. Park Ik-soo, *Founding history of Korean nuclear episodes, 1955 – 1980* (in Korean), Gyeongrim, 2004.
4. Chung Rak-eun, *Status of Nuclear Power in Europe and the US-trip report* (in Korean), Defense Research Institute, Dec. 1958.
5. Lee Kwan, et al., *Feasibility Study Report on Nuclear Power*, NPP-11, Office of Atomic Energy, Dec. 1967.
6. Author interview with Lee Chang-kun.
7. Kim Chan-ok, *Dawn at Sea – Personal Memoir of Kim Chong-joo* (in Korean), pp.162-165, Dosan Mungo, 1996.
8. B&R, *Feasibility Study for the First Nuclear Power Project in the Republic of Korea*, Burns & Roe report to the Office of Atomic Energy, Oct. 31, 1968.
9. Author interview with Kim Duck-seung.
10. D. Harris, I.H. Lee, "Joint Design and Technology Transfer for CANDU Projects in Korea", *presented at the 9th PBNC*, Sydney, Australia, 1994.

Chapter 9
1. Author interview with Shim Chang-saeng.
2. Author interview with Kim Se-jong.
3. Chung Kun-mo, et al, *Design Studies on the Standardization of Nuclear Power Plants*, KOPEC/RR-02/85, vol. 1&2, 1985.

4. Kang, *ibid*, pp. 22-48.
5. Author interview with Park Yong-taek.
6. KOPEC, *30 Year history of Korea Power Engineering*, KOPEC, pp. 49-51, 2005.
7. 7. Author interview with Chung Chung-woon.

Chapter 10
1. Howard, Daniel, *A New History of Old Windsor*, The Journal Press, Windsor Locks, CT, 1935.
2. Author interviews with James Crawford and Tom Natan at Windsor.

Chapter 11
1. Author interview with Regis Matzie at Windsor.
2. Lee Ik-hwan, *Annual Report of Windsor Site Office*, KAERI/GP-62-76/89, internal report, Dec. 1990.
3. Author interview with Jim Veirs at Windsor.
4. Author interview with Jim Veirs at Windsor.
5. Han Ki-in, et al., *Yonggwang 3&4 Final Report on the Reactor System Design Project: Technical Self-reliance Experiences*, KAERI, Dec. 31, 1995.

Chapter 12
1. L.W. Ward, C.P. Fineman, and G.E. Gruen, *INEL Technical Review of YGN 3 and 4 Thermal-Hydraulic Relative Size Effects*, INEL, EGG-2583, August, 1989.
2. Author interview with Chang Soon-heung.
3. Han Ki-in, et al., *Final Report, Yonggwang 3&4 Reactor System Design Project*, Vol. 2, KAERI, 1995.
4. Board of Governors, *Record of the 1112th Meeting*, IAEA GOV/OR.1112, para 98-100, January, 2005.

Chapter 13

1. "License and Technology Transfer Agreement for NSSS System Design and Fuel and Core Design," Contract No: KRI-87-T1112 among KAERI/CE/CEII, clause 7.2 "ROYALTY," 1987.
2. "License Agreement for PWR Technology," Agreement No: KEC-97-T1100 between KEPCO/KHIC/KOPEC/KNFC/KAERI & CE, clause 14.5 "LICENSEE'S Rights," 1997.
3. USDOE, "A Technology Roadmap for Generation IV Nuclear Energy Systems," GIF-002-00, Dec. 2002.
4. Mohammed Al-Fahim, *From Rags to Riches – A story of Abu Dhabi*, The London Centre of Arab Studies, 1995.
5. Roh Eun-rae, "Silk Road and Nuclear Road", *Hydraulic Turbine & Reactor*, KHNP, December, 2005.
6. Kim Kook-hun, "NPP MMIS – from Development to Localization"(in Korean), *Nuclear Industry*, Nov/Dec, 2010.
7. Author interview with Lee Dong-young of KAERI MMIS team.
8. Author interview with Robert Lee of Westinghouse.
9. Hong Nam-pyo, et al., *2010 Nuclear White Paper*, MEST, pp. 119-126, 2010.
10. Author interview with Kim Sun-chang.
11. Author interview with Rieh Chong-hun.

Chapter 14

1. Nuclear Training & Education Center, *Research Reactor Design, Management and Utilization*, KAERI, 2009.
2. Ki Won Lee, et al., "Final Status of the Decommissioning of Research Reactors in Korea," *Journal of Nuclear Science & Technology*, Japan, vol. 47, no. 12, pp. 1227-1232, 2010.
3. Kim Dong-hoon, et al., *HANARO Project Summary Report* (in Korean), KAERI, 1997.
4. Lee Seung-koo, et al., *The Study on Utilization of Senior Nuclear Specialists and their Knowledge Dissemination*, National Research Foundation 20090094175, pp. 84-87, 2010.

Epilogue
1. Economist.com/debate, "*Nuclear power: This house believes that the world would be better off without nuclear power,*" April 15, 2011.

BIBLIOGRAPHY

Following references used in this book are written in Korean only, whose English titles are shown below. For publications with group of authors, the editor-in-chief's name is cited with "ed." after the name.

Cha Jong-hee, *Era of Glory and Indulgence – 30 Years in Nuclear*, autobiography, Shinwoo-sa, Daejeon, 1994.
Cha Jong-hee, ed. *Heavy Water Power Reactor Feasibility Study: Canada Travel Report*, KAERI/PP-74/16, 1973.
Cha Jong-hee, ed. *Heavy Water Power Reactor Feasibility Study*, KAERI/TR/59/78, 1978.
Chang Sun-sup, ed. *White Paper on the LWR Project*, Office of Planning for the LWR Project, 2007.
Chung Kun-mo, ed. *Design Studies on the Standardization of Nuclear Power Plants*, KOPEC/RR-02/85, vol. 1&2, 1985.
Chung Rak-eun, *Status of Nuclear Power in Europe and the US –Trip Report*, Defense Research Institute, Dec., 1958.

Han Ki-in, ed. *Final Report – Yonggwang 3&4 Reactor System Design Project*, KAERI, 1995.
Han Ki-in, ed. *Yonggwang 3&4 Final Report on the Reactor System Design Project: Technical Self-reliance Experiences*, KAERI internal report, Dec. 31, 1995.
Han Pil-soon, "From Hand Grenade to Reactor – Stories to be Told", *Joong-Ang Daily*, 75 column series, Nov. 1999 – April 2000.
Han Pil-soon, *Inauguration Speech as KAERI President*, April 9, 1984.

Han Pil-soon, *Our Choice for 21st Century – Self-reliance of Nuclear Technology*, KAERI, 1997.
Han Pil-soon, ed. *Conditions for an Energy Independent Nation*, KAERI, 1985.
Hong Nam-pyo, ed. *2010 Nuclear White Paper*, MEST, 2010.

KAERI, *NSSS Joint Design Bid Evaluation Report*, YGN internal report, June 30, 1986.
KAERI, *NSSS System Evaluation Report*, YGN internal report, June 30, 1986.
KAERI, *NSSS Technology Transfer Bid Evaluation Report*, YGN internal report, June 30, 1986.
KAERI, *Workshop Proceedings of the Electric Power Group Cooperation Council*, 1986-1997.
Kang Bak-kwang, ed. *A Study on the Effects of '70s~'90s Major Science Policies on S&T and Industrial Development*, MOST Policy Research 2006-21 publication, 2007.
KEPCO, *Meeting Minutes of the 4th EPGCC Main Meeting*, internal document, June 25, 1985.
Kim Chan-ok, *Dawn at Sea – Personal Memoir of Kim Chong-joo*, Dosan Mungo, Seoul, 1996.
Kim Dong-hoon, ed. *HANARO Project Summary Report*, KAERI, 1997.
Kim Dong-hoon, ed. *30th Anniversary Publication –History of KAERI*, KAERI, 1989.
Kim Jong-shin, ed. *Experiences of NPP Technical Self-reliance*, KEPCO, 1996.
Kim Jong-shin, ed. *Wonderful Energy, Beautiful Future – Korean Nuclear Power 30 Year History*, KHNP, 2008.
Kim Seong-hoe, *Forbidden Temptation – Dual Faces of the Atom*, Yewoo Books, Seoul, 2009.
KNFDI, *CANDU-PHWR Nuclear Fuel Localization Plan*, internal report, July 1980.
KNFDI, *PWR Nuclear Fuel Localization Project Plan*, internal report, July 1980.
KOPEC, *30 Year History of Korea Power Engineering*, KOPEC, 2005.

Lee Byung-ryung, *A Study on the Analysis and Countermeasures of Domestic and* Foreign *Impediment Factors for Technical Indigenization*, KAERI/RR-1874/98, 1998.

Lee Byung-ryung, *Korean Light Water Reactor, No Need for US Approval*, Four Seasons, Seoul, 1996.

Lee Chang-kun, *A Vision for the Second Fifty Years of Nuclear Energy – Vision & Strategies*, International Nuclear Societies Council, Oct., 1996.

Lee Hae, ed. *Establishment of Nuclear Engineering Test & Evaluation Center*, KAERI/421/RR-154/80, 1980.

Lee Hae, ed. *Heavy Water Reactor Localization Study: Summary of the JCKS Report*, KAERI/TR/72/78, 1978.

Lee Ik-hwan, *Annual Report of Windsor Site Office*, KAERI/GP-62-76/89, internal report, Dec., 1990.

Lee Jeong-hoon, *Korean Nuclear Sovereignty – Green Growth, still with Nuclear*, Gulmadang, Seoul, 2009.

Lee Kwan, ed. *Feasibility Study Report on Nuclear Power*, NPP-11, Office of Atomic Energy, Dec., 1967.

Lee Seung-hyuk, ed. *Promise & Trust – 20th Anniversary of Korea Institute of Nuclear Safety*, KINS, 2010.

Lee Seung-koo, ed. *The Study on Utilization of Senior Nuclear Specialists and their Knowledge Dissemination*, National Research Foundation 20090094175, 2010.

Lim Yong-kyu, *Seeking Work and Worthiness - My Half Century with Nuclear*, autobiography, Seoul, 2010.

Mok Yong-gyun, ed. *Treatise in Commemoration of Prof. Yoon Young-ku's Retirement*, KAIST, 1994.

Park Ik-soo, *Founding History of Korean Nuclear Episodes, 1955-1980*, Gyeongrim, Seoul, 2004.

Park Ik-soo, *Untold History of Korean Nuclear with Critiques*, Gyeongrim, Seoul, 2004.

Park Jung-ki, *Leadership*, Life & Dream, Seoul, 2002.

Park Jung-ki, *Plain Stories by a Grandpa*, Munmu Publications, Seoul, 1990.

Park Ki-chul, ed. *"50-Year History of Nuclear Korea"*, Korean Nuclear Society, 2010.

Rieh Chong-hun, "Development of Korean Standardized Nuclear Plant and Overseas Export Strategy", *Nuclear Industry*, Nov., 2007.
Rieh Chong-hun, "Stories behind the Early Nuclear Power Planning in Korea", *Hydraulic Turbine & Reactor*, KHNP, July, 2007.
Roh Eun-rae, "Silk Road and Nuclear Road", *Hydraulic Turbine & Reactor*, KHNP, Dec., 2005.

Suh In-suk, ed. *50th Anniversary Publication: History of KAERI*, KAERI, 2009.
Yang Chang-kuk, *Janus' Flame - Story of Nuclear Developments in Two Koreas*, Jigumunhak, Seoul, 2010.

Further information on topics in this all-too-brief history of nuclear power in English could be found in the following references.

Al-Fahim, Mohammed, *From Rags to Riches – A Story of Abu Dhabi*, The London Center of Arab Studies, 1995.

Board of Governors, *Record of the 1112th Meeting*, IAEA GOV/OR.1112, January, 2005.
Burns & Roe, *Feasibility Study for the First Nuclear Power Project in the Republic of Korea*, internal report to the Office of Atomic Energy, Oct. 31, 1968.

Economist.com/debate, *"Nuclear power: This house believes that the world would be better off without nuclear power,"* April 15, 2011.

Harris, D. and Lee Ik-hwan, "Joint Design and Technology Transfer for CANDU Projects in Korea", *presented at the 9th PBNC*, Sydney, Australia, 1994.
Howard, Daniel, *A New History of Old Windsor*, The Journal Press, Windsor Locks, CT, 1935.

KAERI/CE/CEII, *License and Technology Transfer Agreement for NSSS System Design and Fuel and Core Design*, Contract No. KRI-87-T1112, 1987.
KEPCO/KHIC/KOPEC/KNFC/KAERI & CE, *License Agreement for PWR Technology*, Agreement No. KEC-97-T1100, 1997.
Kim Dong-won, "Imaginary Savior: The Image of the Nuclear Bomb in Korea, 1945-1960", *Historia Scientiarum*, 19-2, 2009.

Lee Hae and Pon, George, ed. *Korean CANDU Program: Joint Canada-Korea Study*, KAERI/AECL publication, June 1978.
Lee Ki-won, ed. "Final Status of the Decommissioning of Research Reactors in Korea", *J. of Nuclear Science & Technology*, Japan, vol. 47-12, 2010.
Lee Kwang-seok, ed. *50 Years of Nuclear Energy, 50 Years of Prosperity*, KAERI/MEST, 2009.
Lish, Kenneth C., *Nuclear Power Plant Systems and Equipment*, Industrial Press Inc., 1972.

Mahaffey, James, *Atomic Awakening – A New Look at the History and Future of Nuclear Power*, Pegasus Books, New York, 2009.
Min Byung-joo, ed. *Research Reactor Design, Management and Utilization*, KAERI, 2009.
Murakami, Tomoko, *Commercial War on Nuclear International – Analysis of Market Competitiveness* (in Japanese), Energy Forum, Tokyo, 2010.

Nakamura, Yasuji, *Untold History of Nuclear Fuel Development* (in Japanese), Electric Press, Tokyo, 1990.
Nuclear HRD Center, *Nuclear Power Project: Policy and Korean Experience*, KAERI, First Edition, 2007.

Regulatory Guide 1.20, *Comprehensive Vibration Assessment Program*, USNRC, 1976.
Regulatory Guide 1.155, *Station Blackout*, USNRC, 1988.
Rhee Syngman, *Neutrality as Influenced by the United States*, (PhD dissertation, Princeton University, 1912), Kwanaksa, Seoul, 2004.

Tucker, William, *Terrestrial Energy – How Nuclear Power Will Lead the Green Revolution and End America's Energy Odyssey*, Bartleby Press, Washington, 2008.

USDOE, *A Technology Roadmap for Generation IV Nuclear Energy Systems*, GIF-002-00, Dec., 2002.

Ward, L.W., Fineman, C.P. and Gruen, G.E., *INEL Technical Review of YGN 3&4 Thermal-Hydraulic Relative Size Effects*, INEL EGG-2583, August, 1989.
Ward, Tony, ed. *Nuclear Perspectives – Making tough decisions in a time of pragmatism*, Ernst & Young, 2010.

INDEX

1
123 Agreement, 188

9
95-by-95, 160

A
Abu Dhabi, 186, 189
Additional Protocol, 174, 237
Admiral Hyman Rickover, 25
advanced fuel cycle, 111
Advanced Gas-cooled Reactors (AGR)., 106
Advanced Light Water Reactor (ALWR), 55, 179
Advisory Committee on Reactor Safety (ACRS), 169
AECL, 34, 110
Agency for Defense Development (ADD), 13
Agreed Framework, 91, 94
anti-nuclear movement, 84, 85, 86
AP1000, 106, 142, 185
applied mechanics laboratory, 19
APR+, 195

APR1400, 185, 189, 195
architect engineering company, 54, 62, 123
Areva, 187, 239
Asea Brown Boveri (ABB), 141
ASME N-stamp, 125, 159, 202
Assman, Helmut, 32
ATLAS, 114, 182
Atomic Energy Bureau, 103
Atomic Energy Development Bureau, 1
Atomic Energy Research Institute, 104
atomic machine, 99
AVLIS, 175

B
back-end fuel cycle technology, 13
Baik Young-hak, 15
Barnoski, Mike, 133
Bechtel, 54, 66, 116
Bevilacqua, Frank, 140
Beyond Design Basis Event, 223
Blix, Hans, xix, 90
Blue House, 14, 35, 43
Bosken, Jerome (Sam), 22

Braka Nuclear Power Plant (BNPP), 188, 193
Brewer, Shelby, 67
Burkart, Alex, 24
Burns & Roe, 6, 105, 108
buyers' market, 73

C
Cahn, Herb, 143
Calder Hall, 100, 101, 108
CANDU
    fuel, 22, 33
    -type heavy water reactor, 22, 109, 110
CANFLEX (CANDU Flexible) fuel, 111
Carpentino, Fred, 146
CE Daedeok Office, 154
CE Nuclear, 139
CE's System 80+, 141, 167
Cha In-whan, 128
Cha Jong-hee, 7, 20, 55
Chalon plant, 124
Chang In-soon, 31, 32
Chang Soon-heung, 169
Chang's valve, 168
Changwon Doosan plant, 93, 94, 194
Chattanooga plant, 124, 139
Chernobyl, 72, 113, 224, 226
Chicago Pile, 140
Cho Hee-chul, 80
Cho Kwang-hee, xxii
Choi Chang-nak, 78
Choi Chang-oong, xxii
Choi Han-kwon, 94
Choi Suhn, xxii
Choi Yeol, 88
Choi Yeong-sang, 212
Choi Young-jin, 174
Chun Doo-hwan, 1, 26, 35, 75
Chung Chung-woon, 125
Chung In-yung, 124
Chung Ju-yung, 124
Chung Kun-mo, 58, 123
Chung Rak-eun, 103
Chung Sang-myung, 80
Cisler, Walker Lee, 101, 102
Cohen, Kenneth, 22
cold neutron, 210
Colton, John (Phil), 22
Combustion Engineering, 62, 66, 97, 108
component approach, 116, 201
Comprehensive Vibration Assessment Program (CVAP), 167
Construction Permit, 93, 165, 186
control element drive mechanism, 125
core damage frequency, 221
Core Operating Limit Supervisory System (COLSS), 141
Core Protection Calculator (CPC), 141
Country Program Framework (CPF), 69
Crawford, James, 137, 142, 154
Criteria Committee, 40

Crosher, Kenneth, 22
crossing the nuclear Rubicon, 59
Crump, Mark, 17, 137

D
Daedeok Engineering Center, 30, 55
Daedeok Innopolis, 17, 18, 183
Daedeok Science Town, 16, 42, 183
Daedeok Valley, 183
Delivery warranty, 71
desalination plants, 126
design basis earthquake, 221
Design Certificate, 181, 186
Design Document Control Center (DDCC), 154, 159
Design Technology Independence, 131
Development Assistance Committee, 214
Direct Vessel Injection (DVI), 182
Doosan Heavy Industries, 126, 240
Doryong-dong, 43, 44
dual-use technology, 22
Duncan, Murray, 18
DUPIC, 110

E
Economic Planning Board, 15, 107
Eisenhower's 'Atoms for Peace', 100
Electric Power Group Cooperation Council, 58, 121, 161

Electric Power Research Institute (EPRI), 55, 179
Emirates Nuclear Energy Corporation (ENEC), 188
Enertopia, 48
Engineering Logic Network, 150
Evaluation Committee, 40
export control, 170

F
fast breeder technology, 45
Federal Authority for Nuclear Regulation (FANR), 188
feed-and-bleed system, 168
Final Design Approval, 78
Final Safety Analysis Report (FSAR), 156
first concrete, 93, 165, 193
first oil crisis, 104
fluid systems design, 146, 152
fluidic device, 182, 186
Framatome, 52, 65, 125
French EPR technology, 52
French Loan Projects, 8
Fukushima Daiichi, 220, 221
full-scope safeguards agreement, 173
full-scope technology transfer, 122, 143

G
General Terms & Conditions, 71
Generation I NPPs, 185
Generation II NPPs, 160, 185
Generation III, 181, 185

Generation IV reactor technology, 89
Gill, Bill, 147
Gonzales, Abel, 18
Grant of License, 197, 198
Gulf Cooperation Council (GCC), 189

H
Hahm Chang-shik, 61
Han Jae-bok, 146
Han Ki-in, xxii, 127
Han Pil-soon, 44, 47, 86, 127, 187
HANARO, 209, 210
Hanyang University, 5, 133, 206
Hart, Robert, 35
high-level wastes, 87
history of Windsor, 138
Hitachi-GE, 64, 186
Hong Young-soo, 125
hot functional testing, 167
hot test loop, 33, 34, 182
hydrogen explosion, 222, 224
Hyun Kyung-ho, 7
Hyundai Construction, 128
Hyundai International, 124

I
Indian nuclear test, 8
initial core design, 71, 145, 151, 181
in-pile testing, 34
In-reactor Water Storage Tank (IRWST), 182

Institute of Nuclear Materials Management (INMM), 173
Institute of Nuclear Power Operators, 19
International Atomic Energy Agency, 19, 90, 173
International Nuclear and Radiological Event Scale, 222
International Nuclear Safety Advisory Group (INSAG), 169
INVAP, 210
invitation-to-bid, 40, 62, 64

J
Jeon Jae-poong, xxii
Joint Canada-Korea Study, 7
Joint Declaration of Denuclearization, 9
Joint Design, 145
Joint Standing Committee on Nuclear and Other Energy Technologies, 23
Joint System Design, 71, 150
Jordan Research and Training Reactor (JRTR), 210

K
KABAR, 6, 123
KAERI
    close down, 1
    daughter companies of, 126
    in the construction project, xxvii
    of power projects, 56

radwaste site program out of, 86, 87
recommendation on NSSS vendor, 68, 79
the national nuclear laboratory, 18, 19
Windsor Office, 134, 150

KAIST, 1, 18, 182
Kang Bak-kwang, 1, 81
KEDO LWR project, 91, 93
KEPCO
consortium, 94, 186, 200
daughter company of, 126, 127
first export project, 186
prime contractor for NPP contract, 70, 71
project review meetings (PRM) called by, 120
top management, 65
total project management, 122

Khalifa University, 189
KHIC, 59, 70, 71, 124
Kim Byung-koo, 31, 61, 212
Kim Chong-joo, 6, 107
Kim Dae-jung, 82
Kim Dong-hoon, 207, 208
Kim Dong-soo, 146
Kim Duck-seung, 109
Kim Il-Sung, 100
Kim Jae-ik, 43
Kim Jae-poong, 127
Kim Jin-soo, 81, 133

Kim Jong-shik, 166
Kim Kook-hun, 196
Kim Se-jong, 119
Kim Seong-yun, 31, 211
Kim Si-hwan, 39, 61, 200
Kim Sun-chang, 39, 58
Kim Sung-jin, 26, 28
Kim Young-june, 3, 201
KINAC, 18, 176, 184
KINS, 164, 165, 184, 195
KIRAMS, 126
Kirby, Robert, 78
Kirkman, Robert, 155
Koh Byung-joon, 165
Koh Joong-myung, 187
Korea Advanced Energy Research Institute, 21
Korea Air Force Academy, 29
Korea Electric Company (KECO), 52, 54
Korea Electric Power Corporation (KEPCO), 186
Korea Electric Power Industry Codes (KEPIC), 202
Korea Electric Power Research Institute (KEPRI), 184
Korea Hydro and Nuclear Power (KHNP), 182, 184, 186
Korea Military Academy, 28, 48
Korea Multipurpose Research Reactor(KMRR), 209
Korea Next-Generation Reactor (KNGR), 180, 182, 184
Korea Nuclear Fuel Company, 10, 70

Korea Nuclear Fuel Development Institute, 9, 10, 30
Korea Nuclear Society, 18
Korea Power Engineering Company (KOPEC), 57, 124, 223
Korea Radioactive-waste Management Corporation, 8
Korean Pilgrims, 129
Korean Standardized Nuclear Plants, 118, 178
Kori Unit 1, 177, 210
Krecicki, Vince, 143
K-STAR, 18
Kuh Jung-eui, 146
Kumho site, 93, 112
Kwon Joong-gyu, 125
Kwon Kee-choon, 196
KWU, 40, 62, 145

L

Least Developed Country, 214
Lee, Benjamin Whisoh, 21
Lee Byung-ryung, 61, 146, 212
Lee Byung-whie, 45
Lee Chang-kun, 66, 202
Lee Cheong-o, 1
Lee Gyu-am, 31
Lee Hae, xxii, 6, 7
Lee Ik-hwan, xxii, 61, 133
Lee Kwan, 105
Lee Myung-bak, 128
Lee, Robert S., xxii, 197
Lee Sang-hoon, 4, 165
Lee Seung-hyuk, 165
Lee Seung-koo, xxii

Licensing Agreement, 178
Lichtenberger, Harold, 140
Liimatainen, Robert, 22
Lim Han-kwae, xxii, 70
Longo, Joe, 146
low/medium-level wastes, 87, 88
LWR fuel localization project, 37
Lyu Jun-sang, 81

M

Magnox-type gas-cooled reactor, 63, 106
Man Machine Interface System (MMIS), 195
MAPLE-X reactors, 209
Markovitz, Karl, 143
Matzie, Regis, 137, 147, 170
mechanical design and analysis (MDA), 146
medical/industrial isotopes, 205
Michio, Ishikawa, 18
Ministry of Education, Science & Technology (MEST), 195
Ministry of Energy & Resources, 3, 15
Ministry of Science & Technology, 1, 185
Mokpo Declaration, 82
Moon Hee-sung, 7
mutual inspection, 172, 173

N

Nakamura Sensei, 44
Nam Jang-soo, xxiii, 14, 31, 40
Natan, Tom, xxii, 137, 142, 156

National Assembly audit, 166
national regulatory authority, 164
National Research Foundation, 18
natural uranium fuel, 110
Nautilus nuclear submarine, 117
newcomer countries, 216
Newman, Bob, 143, 153
next-generation reactor development, 179
Niels Bohr Institute, 43
NIMBY, 84, 87
North Korean nuclear crises, 91, 94
North-South relationship, 95
NPP standardization, 53, 118, 119
NPP technical self-reliance, 97, 120, 121, 181
NRU reactor, 34
NSSS system design, 56, 61, 71, 79
Nuclear Control Joint Committee (NCJC), 171
nuclear dark age, 113, 128
Nuclear engineering departments, 206
Nuclear Engineering Documentation System (NEDS), 153
nuclear material accountancy control, 171, 172
nuclear non-proliferation agency, 175
Nuclear Nonproliferation Treaty, 8, 173
nuclear renaissance, 216, 226
Nuclear Research Fund, 181
Nuclear Safety Center, 14, 164
Nuclear Security Summit, 176
Nuclear Silk Road, 217
Nuclear Technology Mecca, 18, 181
Nuclear Valley, 17, 183
nuclear-grade components, 121

O

Office of Atomic Energy, 104
Operation License, 164, 170
Optimized Power Reactor (OPR), 178
Out-of-pile testing, 31, 34

P

Palo Verde NPP, 166, 170
Park Chung-hee, 26, 28, 75
Park Ik-soo, 103
Park Jung-ki, xxii, 48, 49
Park Yong-taek, xxii, 122, 170
passive safety features, 182, 186
Peck, Dan, 147, 156
performance warranty, 39, 71, 157
physical protection, 176
Pon, George, 7
post-irradiation examination facility, 8
Power Reactor Systems Division, 61, 158
pressurized water reactor (PWR), 11, 106, 109
pressurizer relief valve, 168
probabilistic safety analysis, 170
Program 93+2, 173
proliferation-resistant, 89

prosecutor's investigation, 80, 81
PSAR, 18, 154, 186
Pucak, Jack, 155
Purex process, 89
pyro-processing, 89

R
Radiation Agricultural Institute, 5
Radiation Medical Institute, 5
radwaste disposal program, 86
radwaste disposal site, 86, 87
Radwaste Fund, 86
reactor coolant loop, 51, 199
reactor coolant pumps, 194, 199
reload fuel design, 145
reverse engineering, 34
Rhee Syngman, xxv, 99, 102
Rieh Chong-hun, xxii, 122, 212
Rim Chang-saeng, xxii, 39, 61
Roh Eun-rae, xxii, 7, 194
Roh Moo-hyun, 87
Roh Tae-woo, 75
Rosen, Morris, 18
royalty payment, 197
royalty-free use rights, 72, 73, 198
Russian VVER, 93

S
Safe Shutdown Earthquake, 180
safety culture, 227
Safety Depressurization System, 168
Safety Evaluation Report, 195
SAGSI, 173

Sargent & Lundy, 86, 115, 120
scaled-down hybrid, 78, 166
Seoul National University, 5, 206
Seoul Olympics, 75, 84, 163
Sessom, Alan, 23
severe accident, 168, 182, 224
Sheikh Zayed, 190
Shilla golden crown, 148
Shim Chang-saeng, xxii, 70, 122, 187
Shin Hyun-kook, 133
Shin Jae-in, 119
Shin-Kori Units 3&4, 177
Shippingport, 185
Silk Road, 216, 217
Six-Party Talks, 91, 95
slightly enriched uranium, 110
small and medium reactor (SMR), 199
SMART, 198
Sohn Gap-heon, xxii, 61, 146
Sovka, Jerry, 33
Standard Design Approval, 181, 184, 200
Standard Safety Analysis Report, 186
State System of Accountancy Control (SSAC), 173
station blackout (SBO), 221, 223
Stella, Robert, 13, 22
Suh Kyung-soo, 31, 36
Suh Sang-cheol, 118
Suh Suk-chun, xxii, 32
Suk Ho-chun, 31

Sung Nak-jung, 58
System 80 design, 78, 167
System 80+, 142
System Design Support, 158

T
Tae Wan-sun, 107
Taiwan Research Reactor, 13
tandem fuel cycle, 110
Technology Center for Nuclear Control (TCNC), 173
technology over politics, 1, 67
Technology Transfer contract, 71
ten-minute principle, 43
The Big One, 220
thermal/hydraulic relative size, 166
thermal/hydraulic safety, 167
thermal-hydraulic laboratory, 19
TMI accident, 168, 179, 224
TMI Action Items, 179
Tokai Reprocessing Plant, 44
Tokyo Electric Power Company (TEPCO), 220
Topical Reports, 186, 200
Toshiba/Westinghouse merger, 142
Transition Plan, 151, 154, 156, 159
TRIGA Mark-II, 5, 100, 206
tsunami, 220, 221
turnkey project, 93, 110

U
UAE, 137, 188, 190, 216
Ulchin 3&4 project, 92, 113, 158, 177
unfinished project, 94
uranium conversion plants, 32
uranium enrichment program, 94
uranium oxide powder, 32
uranium refining, 16
US Navy submarine propulsion system, 25
US-Japan Nuclear Cooperation Agreement, 45
USNRC, 166, 181, 189
utility requirements document, 179

V
Veirs, Jim, xxii, 70, 133, 137, 143
verification testing, 34, 195

W
West, John, 140
Westinghouse, 52, 66, 78, 116, 201
Wolsong fuel localization project, 33, 36
Wolsong Unit 1, 9, 31

Y
Yang Chang-kook, 40
Yang Seung-yeong, 61
Yongbyon, 90, 172
Yonggwang 3&4, 62, 160
Yoo Joo-young, 124
Yoon Se-won, 103
Yoon Young-ku, xxii, 5

Z
Zech, Lando, 166
Zinn, Walter, 140

Made in the USA
Lexington, KY
03 May 2012